Computer Vision on AWS

Build and deploy real-world CV solutions with Amazon
Rekognition, Lookout for Vision, and SageMaker

Lauren Mullennex

Nate Bachmeier

Jay Rao

BIRMINGHAM—MUMBAI

Computer Vision on AWS

Publishing Product Manager: Dinesh Chaudhary
Content Development Editor: Joseph Sunil
Technical Editor: Sweety Pagaria
Copy Editor: Safis Editing
Project Coordinator: Farheen Fathima
Proofreader: Safis Editing
Indexer: Manju Arasan
Production Designer: Ponraj Dhandapani
Marketing Coordinator: Shifa Ansari

First published: April 2023

Production reference: 1240323

Published by Packt Publishing Ltd.
Livery Place
35 Livery Street
Birmingham
B3 2PB, UK.

ISBN 978-1-80107-868-9

www.packtpub.com

To my father, brother, mother, and Denise – there are not enough words to express how grateful I am for your support and guidance. Thank you for your encouragement and for teaching me how to persevere.

– Lauren Mullennex

To my parents, sister, and my wife – who always believed in me. Thank you for your continued love and support!

– Jay Rao

Contributors

About the authors

Lauren Mullennex is a senior AI/ML specialist solutions architect at AWS. She has broad experience in infrastructure, DevOps, and cloud architecture across multiple industries. She has published multiple AWS AI/ML blogs, spoken at AWS conferences, and focuses on developing solutions using CV and MLOps.

Nate Bachmeier is a principal solutions architect at AWS (PhD. CS, MBA). He nomadically explores the world one cloud integration at a time, focusing on the financial service industry.

Jay Rao is a principal solutions architect at AWS. He enjoys providing technical and strategic guidance to customers and helping them design and implement solutions.

About the reviewer

Morteza Kiadi is a seasoned cloud computing and machine learning technical instructor with over fifteen years of expertise. Morteza pursued his Ph.D. in AI and Optimization while held several positions in enterprises and startups. As a senior technical trainer at AWS, Morteza instructs all machine learning courses in AWS, supporting AWS customers in mastering challenging topics and cutting-edge technology. Morteza utilized his skills in academia, industry, teaching, and analyzing other publications to render this book accessible to its readers by providing insightful and impartial reviews.

I am extremely grateful to Emily Webber for inspiring me to review this book. I am thankful to the entire Packt Publishing team for your ongoing assistance in the publishing of this book.

Table of Contents

3

Creating Custom Models with Amazon Rekognition Custom Labels   41

Part 2: Applying CV to Real-World Use Cases

4

Using Identity Verification to Build a Contactless Hotel Check-In System   61

5

Automating a Video Analysis Pipeline   85

6

Moderating Content with AWS AI Services   115

Part 3: CV at the edge

7

Introducing Amazon Lookout for Vision   143

8

Detecting Manufacturing Defects Using CV at the Edge   165

Part 4: Building CV Solutions with Amazon SageMaker

9

10

Part 5: Best Practices for Production-Ready CV Workloads

11

12

13

Applying AI Governance in CV   275

Index   287

Other Books You May Enjoy   300

Preface

Computer vision (**CV**) transforms visual data into actionable insights to solve many business challenges. In recent years, due to the availability of increased computing power and access to vast amounts of data, CV has become more accessible. **Amazon Web Services** (**AWS**) has played an important role in democratizing CV by providing services to build, train, and deploy CV models.

In this book, you will begin by exploring the applications of CV and features of Amazon Rekognition and Amazon Lookout for Vision. Then, you'll walk through real-world use cases such as identity verification, real-time video analysis, content moderation, and detecting manufacturing defects to understand how to implement AWS AI/ML services. You'll also use Amazon SageMaker for data annotation, training, and deploying CV models. As you progress, you'll learn best practices and design principles for scaling, reducing cost, improving the security posture, and mitigating the bias of CV workloads.

By the end of this book, you'll be able to accelerate your business outcomes by building and implementing CV into your production environments with AWS AI/ML services.

Who this book is for

If you are a machine learning engineer, a data scientist, or want to better understand best practices and how to build comprehensive CV solutions on AWS, this book is for you. Knowledge of AWS basics is required to grasp the concepts covered in this book more effectively. A solid understanding of ML concepts and the Python programming language will also be beneficial.

What this book covers

Chapter 1, Computer Vision Applications and AWS AI/ML Overview, provides an introduction to CV and summarizes use cases where CV can be applied to solve business challenges. It also includes an overview of the AWS AI/ML services.

Chapter 2, Interacting with Amazon Rekognition, covers an overview of Amazon Rekognition and details the different capabilities available, including walking through the Amazon Rekognition console, and how to use the APIs.

Chapter 3, Creating Custom Models with Amazon Rekognition Custom Labels, provides a detailed introduction to Amazon Rekognition Custom Labels, what its benefits are, and a code example to train a custom object detection model.

Chapter 4, Using Identity Verification to Build a Contactless Hotel Check-In System, dives deep into a real-world use case using Amazon Rekognition and other AWS AI services to build applications that demonstrate how to solve business challenges using core CV capabilities. A code example is provided to build a mobile application for customers to register their faces and check into a fictional hotel kiosk system.

Chapter 5, Automating a Video Analysis Pipeline, dives deep into a real-world use case using Amazon Rekognition to build an application that demonstrates how to solve business challenges using core CV capabilities. A code example is provided to build a real-time video analysis pipeline using Amazon Rekognition Video APIs.

Chapter 6, Moderating Content with AWS AI Services, dives deep into a real-world use case using Amazon Rekognition and other AWS AI services to build applications that demonstrate how to solve business challenges using core CV capabilities. A code example is provided to build content moderation workflows.

Chapter 7, Introducing Amazon Lookout for Computer Vision, provides a detailed introduction to Amazon Lookout for Vision, what its functions are, and a code example to train a model to detect anomalies.

Chapter 8, Detecting Manufacturing Defects Using CV at the Edge, dives deeper into Amazon Lookout for Vision, covers the benefits of deploying CV at the edge, and walks through a code example to train a model to detect anomalies in manufacturing parts.

Chapter 9, Labeling Data with Amazon SageMaker Ground Truth, provides a detailed introduction to Amazon SageMaker Ground Truth, what its benefits are, and a code example to integrate a human labeling job into offline data labeling workflows.

Chapter 10, Using Amazon SageMaker for ComputerVision, dives deeper into Amazon SageMaker, covers its capabilities, and walks through a code example to train a model using a built-in image classifier.

Chapter 11, Integrating Human-in-the-Loop with Amazon Augmented AI, provides a detailed introduction to Amazon Augmented AI (Amazon A2I), what its functions are, and a code example that uses human reviewers to improve the accuracy of your CV workflows.

Chapter 12, Best Practices for Designing an End-to-End CV Pipeline, covers best practices that can be applied to CV workloads across the entire ML lifecycle, including considerations for cost optimization, scaling, security, and developing an MLOps strategy.

Chapter 13, Applying AI Governance in CV, discusses the purpose of establishing an AI governance framework, introduces Amazon SageMaker for ML governance, and provides an overview of the importance of mitigating bias.

To get the most out of this book

You will need access to an AWS account, so before getting started, we recommend that you create one.

Software/hardware covered in the book	Operating system requirements/Account creation requirements
Access to or signing up for an AWS account	`https://portal.aws.amazon.com/billing/signup`
Jupyter Notebook	Windows/macOS

If you are using the digital version of this book, we advise you to type the code yourself or access the code from the book's GitHub repository (a link is available in the next section). Doing so will help you avoid any potential errors related to the copying and pasting of code.

Download the example code files

You can download the example code files for this book from GitHub at `https://github.com/PacktPublishing/Computer-Vision-on-AWS`. If there's an update to the code, it will be updated in the GitHub repository.

We also have other code bundles from our rich catalog of books and videos available at `https://github.com/PacktPublishing/`. Check them out!

Conventions used

There are a number of text conventions used throughout this book.

`Code in text`: Indicates code words in text, database table names, folder names, filenames, file extensions, pathnames, dummy URLs, user input, and Twitter handles. Here is an example: "Once the model is hosted, you can start analyzing your images using the `DetectAnomalies` API."

A block of code is set as follows:

```
{
    "SubscriptionArn": "arn:aws:sns:region:account:AmazonRekogn
itionPersonTrackingTopic:04877b15-7c19-4ce5-b958-969c5b9a1ecb"
}
```

When we wish to draw your attention to a particular part of a code block, the relevant lines or items are set in bold:

```
[aws sns subscribe \
  --region us-east-2 \
  --topic-arn
arn:aws:sns:region:account:AmazonRekognitionPersonTrackingTopic
  \
```

```
  --protocol sqs \
  --notification-endpoint
arn:aws:sqs:region:account:PersonTrackingQueue
```

Any command-line input or output is written as follows:

```
$ git clone https://github.com/PacktPublishing/Computer-Vision-
on-AWS
$ cd Computer-Vision-on-AWS/07_LookoutForVision
```

Bold: Indicates a new term, an important word, or words that you see onscreen. For instance, words in menus or dialog boxes appear in **bold**. Here is an example: "If you're a first-time user of the service, it will ask permission to create an S3 bucket to store your project files. Click **Create S3 bucket**."

> **Tips or important notes**
> Appear like this.

Get in touch

Feedback from our readers is always welcome.

General feedback: If you have questions about any aspect of this book, email us at customercare@ packtpub.com and mention the book title in the subject of your message.

Errata: Although we have taken every care to ensure the accuracy of our content, mistakes do happen. If you have found a mistake in this book, we would be grateful if you would report this to us. Please visit www.packtpub.com/support/errata and fill in the form.

Piracy: If you come across any illegal copies of our works in any form on the internet, we would be grateful if you would provide us with the location address or website name. Please contact us at copyright@packt.com with a link to the material.

If you are interested in becoming an author: If there is a topic that you have expertise in and you are interested in either writing or contributing to a book, please visit authors.packtpub.com.

Share Your Thoughts

Once you've read *Computer Vision on AWS*, we'd love to hear your thoughts! Scan the QR code below to go straight to the Amazon review page for this book and share your feedback.

https://packt.link/r/1-801-07868-8

Your review is important to us and the tech community and will help us make sure we're delivering excellent quality content.

Download a free PDF copy of this book

Thanks for purchasing this book!

Do you like to read on the go but are unable to carry your print books everywhere?

Is your eBook purchase not compatible with the device of your choice?

Don't worry, now with every Packt book you get a DRM-free PDF version of that book at no cost.

Read anywhere, any place, on any device. Search, copy, and paste code from your favorite technical books directly into your application.

The perks don't stop there, you can get exclusive access to discounts, newsletters, and great free content in your inbox daily

Follow these simple steps to get the benefits:

1. Scan the QR code or visit the link below

https://packt.link/free-ebook/9781801078689

2. Submit your proof of purchase
3. That's it! We'll send your free PDF and other benefits to your email directly

Part 1: Introduction to CV on AWS and Amazon Rekognition

As a machine learning engineer or data scientist, this section helps you better understand how CV can be applied to solve business challenges and gives a comprehensive overview of the AWS AI/ML services available.

This first part consists of three cumulative chapters that will cover the core concepts of CV, a detailed introduction to Amazon Rekognition, and how to create a custom classification model using Amazon Rekognition Custom Labels.

By the end of this part, you will understand how to apply CV to accelerate your business outcomes, what AWS AI/ML services for your CV workloads, and how to use Amazon Rekognition for tasks including classification and object detection.

This part comprises the following chapters:

- *Chapter 1, Computer Vision Applications and AWS AI/ML Overview*
- *Chapter 2, Interacting with Amazon Rekognition*
- *Chapter 3, Creating Custom Models with Amazon Rekognition Custom Labels*

Computer Vision Applications and AWS AI/ML Services Overview

In the past decade, the field of **computer vision** (**CV**) has rapidly advanced. Research in **deep learning** (**DL**) techniques has helped computers mimic human brains to "see" content in videos and images and transform it into actionable insights. There are examples of the wide variety of applications of CV all around us, including self-driving cars, text and handwriting detection, classifying different types of skin cancer in images, industrial equipment inspection, and detecting faces and objects in videos. Despite recent advancements, the availability of vast amounts of data from disparate sources has posed challenges in creating scalable CV solutions that achieve high-quality results. Automating a production CV pipeline is a cumbersome task requiring many steps. You may be asking, "How do I get started?" and "What are the best practices?".

If you are a **machine learning** (**ML**) engineer or data scientist or want to better understand how to build and implement comprehensive CV solutions on **Amazon Web Services** (**AWS**), this book is for you. We provide practical code examples, tips, and step-by-step explanations to help you quickly deploy and automate production CV models. We assume that you have intermediate-level knowledge of **artificial intelligence** (**AI**) and ML concepts. In this first chapter, we will introduce CV and address implementation challenges, discuss the prevalence of CV across a variety of use cases, and learn about AWS AI/ML services.

In this chapter, we will cover the following:

- Understanding CV
- Solving business challenges with CV
- Exploring AWS AI/ML services
- Setting up your AWS environment

Technical requirements

You will need a computer with internet access to create an AWS account to set up Amazon SageMaker to run the code samples in the following chapters. The Python code and sample datasets for the solutions discussed are available at https://github.com/PacktPublishing/Computer-Vision-on-AWS.

Understanding CV

CV is a domain within AI and ML. It enables computers to detect and understand visual inputs (videos and images) to make predictions:

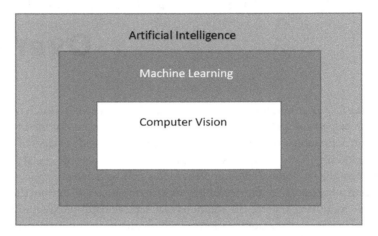

Figure 1.1 – CV is a subdomain of AI and ML

Before we discuss the inner workings of a CV system, let's summarize the different types of ML algorithms:

- **Supervised learning (SL)**—Takes a set of labeled input data and predicts a known target value. For example, a model may be trained on a set of labeled dog images. When a new unlabeled dog image is processed by the model, the model correctly predicts that the image is a dog instead of a cat.

- **Unsupervised learning (UL)**—Unlabeled data is provided, and patterns or structures need to be found within the data since no labeled target value is present. One example of UL is a targeted marketing campaign where customers need to be segmented into groups based on various common attributes such as demographics.

- **Semi-supervised learning**—consists of unlabeled and labeled data. This is beneficial for CV tasks, since it is a time-consuming process to label individual images. With this method, only some of the images in the dataset need to be labeled, in order to label and classify the unlabeled images.

CV architecture and applications

Now that we've covered the different types of ML training methods, how does this relate to CV? DL algorithms are commonly used to solve CV problems. These algorithms are composed of **artificial neural networks (ANNs)** containing layers of nodes, which function like a neuron in a human brain. A **neural network (NN)** has multiple layers, including one or more input layers, hidden layers, and output layers. Input data flows through the input layers. The nodes perform transformations of the input data in the hidden layers and produce output to the output layer. The output layer is where predictions of the input data occur. The following figure shows an example of a **deep NN (DNN)** architecture:

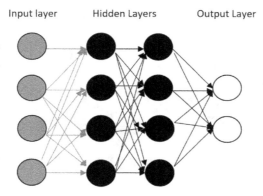

Figure 1.2 – DNN architecture

How does this architecture apply to real-world applications? With CV and DL technology, you can detect patterns in images and use these patterns for classification. One type of NN that excels in classifying images is a **convolutional NN (CNN)**. CNNs were inspired by ANNs. The way the nodes in a CNN communicate replicates how animals visualize the world. One application of CNNs is classifying X-ray images to assist doctors with medical diagnoses:

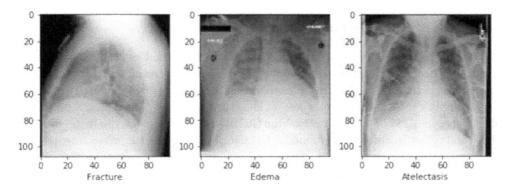

Figure 1.3 – Image classification of X-rays

There are multiple types of problems that CV can solve that we will highlight throughout this book. **Localization** locates one or more objects in an image and draws a bounding box around the object(s). **Object detection** uses localization and classification to identify and classify one or multiple objects in an image. These tasks are more complicated than image classification. **Faster R-CNN (Regions with CNN)**, **SSD (single shot detector)**, and **YOLO (you only look once)** are other types of DNN models that can be used for object detection tasks. These models are designed for performance such as decreasing latency and increasing accuracy.

Segmentation—including instance segmentation and semantic segmentation—highlights the pixels of an image, instead of objects, and classifies them. Segmentation can also be applied to videos to detect black frames, color bars, end credits, and shot changes:

Figure 1.4 – Examples of different CV problem types

Despite recent advances in CV and DL, there are still challenges within the field. CV systems are complex, there are vast amounts of data to process, and considerations need to be taken before training a model. It is important to understand the data available since a model is only as good as the quality of your data, and the steps required to prepare the data for model training.

Data processing and feature engineering

CV deals with images and videos, which are a form of unstructured data. Unstructured data does not have a predefined data model and cannot be stored in a database row and column format. This type of data poses unique challenges compared to tabular data. More processing is required to transform the data into a usable format. A computer sees an image as a matrix of pixel values. A pixel is a set of numbers between 0-255 in the **red, green, blue (RGB)** system. Images vary in their resolutions, dimensions, and colors. In order to train a model, CV algorithms require that images are normalized such that they are the same size. Additional image processing techniques include resizing, rotating, enhancing the resolution, and converting from RGB to grayscale. Another technique is image masking, which allows us to focus on a region of interest. In the following photos, we apply a mask to highlight the motorcycle:

Original image   Applying mask with color overlay

Figure 1.5 – Applying an image mask to highlight the motorcycle

Preprocessing is important since images are often large and take up lots of storage. Resizing an image and converting it to grayscale can speed up the ML training process. However, this technique is not always optimal for the problem we're trying to solve. For example, in medical image analysis such as skin cancer diagnosis, the colors of the samples are relevant for a proper diagnosis. This is why it's important to have a complete understanding of the business problem you're trying to solve before choosing how to process your data. In the following chapters, we'll provide code examples that detail various image preprocessing steps.

Features or attributes in ML are important input data characteristics that affect the output or target variable of a model. Distinct features in an image help a model differentiate objects from one another. Determining relevant features depends on the context of your business problem. If you're trying to identify a Golden Retriever dog in a group of images also containing cats, then height is an important feature. However, if you're looking to classify different types of dogs, then height is not always a distinguishing feature since Golden Retrievers are similar in height to many other dog breeds. In this case, color and coat length might be more useful features.

Data labeling

Data annotation or data labeling is the process of labeling your input datasets. It helps derive value from your unstructured data for SL. Some of the challenges with data labeling are that it is a manual process that is time-consuming, humans have a bias for labeling an object, and it's difficult to scale. **Amazon SageMaker Ground Truth Plus** (https://aws.amazon.com/sagemaker/data-labeling/) helps address these challenges by automating this process. It contains a labeling **user interface (UI)** and quality workflow customizations. The labeling is done by an expert workforce with domain knowledge of the ML tasks to complete. This improves the label quality and leads to better training datasets. In *Chapter 9*, we will cover a code example using SageMaker Ground Truth Plus.

Amazon Rekognition Custom Labels (`https://aws.amazon.com/rekognition/custom-labels-features/`) also provides a visual interface to label your images. Labels can be applied to the entire image or you can create bounding boxes to label specific objects. In the next two chapters, we will discuss Amazon Rekognition and Rekognition Custom Labels in more detail.

In this section, we discussed the architecture behind DL CV algorithms. We also covered data processing, feature engineering, and data labeling considerations to create high-quality training datasets. In the next section, we will discuss the evolution of CV and how it can be applied to many different business use cases.

Solving business challenges with CV

CV has tremendous business value across a variety of industries and use cases. There have also been recent technological advancements that are generating excitement within the field. The first use case of CV was noted over 60 years ago when a digital scanner was used to transform images into grids of numbers. Today, **vision transformers** and **generative AI** allow us to quickly create images and videos from text prompts. The applications of CV are evident across every industry, including healthcare, manufacturing, media and entertainment, retail, agriculture, sports, education, and transportation. Deriving meaningful insights from images and videos has helped accelerate business efficiency and improved the customer experience. In this section, we will briefly cover the latest CV implementations and highlight use cases that we will be diving deeper into throughout this book.

New applications of CV

In 1961, Lawrence Roberts, who is often considered the "father" of CV, presented in his paper *Machine Perception of Three-Dimensional Solids* (`https://dspace.mit.edu/bitstream/handle/1721.1/11589/33959125-MIT.pdf`) how a computer could construct a 3D array of objects from a 2D photograph. This groundbreaking paper led researchers to explore the value of image recognition and object detection. Since the discovery of NNs and DL, the field of CV has made great strides in developing more accurate and efficient models. Earlier, we reviewed some of these models, such as CNN and YOLO. These models are widely adopted for a variety of CV tasks. Recently, a new model called vision transformers has emerged that outperforms CNN in terms of accuracy and efficiency. Before we review vision transformers in more detail, let's summarize the idea of **transformers** and their relevance in CV.

In order to understand transformers, we first need to explore a DL concept that is used in **natural language processing** (**NLP**), called **attention**. An introduction to transformers and self-attention was first presented in the paper *Attention is All You Need* (`https://arxiv.org/pdf/1706.03762.pdf`). The attention mechanism is used in RNN **sequence-to-sequence** (**seq2seq**) models. One example of an application of seq2seq models is language translation. This model is composed of an encoder and a decoder. The encoder processes the input sequence, and the decoder generates the transformed output. There are hidden state vectors that take the input sequence and the context vector from the encoder and send them to the decoder to predict the output sequence. The following diagram is an illustration of these concepts:

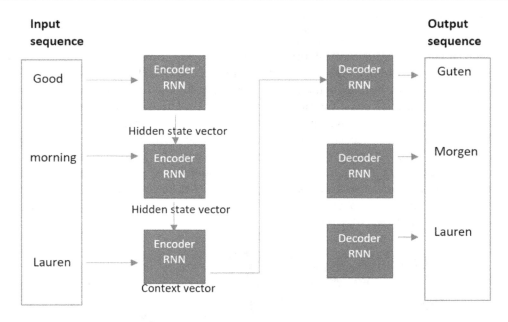

Input sequence

Good

Encoder RNN

Hidden state vector

morning

Encoder RNN

Hidden state vector

Lauren

Encoder RNN

Context vector

Decoder RNN

Decoder RNN

Decoder RNN

Output sequence

Guten

Morgen

Lauren

Figure 1.6 – Translating a sentence from English to German using a seq2seq model

In the above, we pay attention to the context of the words in the input to determine the next sequence when generating the output. Another example of attention from *Attention is All You Need* weighs the importance of different inputs when making predictions. Here is a sentiment analysis example from the paper for a hotel service task, where the bold words are considered relevant:

you get **what** you **pay** for . not the cleanest rooms but bed was **clean** and so was **bathroom** . bring your own **towels** though as very **thin** . **service was excellent** , let us book in at 8:30am ! for location and price , this ca n't be beaten , but it is **cheap** for a reason . if you come expecting the hilton , then book the hilton ! for uk travellers , think of a blackpool b&b.

Figure 1.7 – Example of attention for sentiment analysis from "Attention is All You Need"

A transformer relies on self-attention, which is defined in the paper as "*an attention mechanism relating different positions of a single sequence in order to compute a representation of the sequence*". Transformers are important in the application of NLP because they capture the relationship and context of words in text. Take a look at the following sentences:

Andy Jassy is the current CEO of Amazon. He was previously the CEO of Amazon Web Services.

Using transformers, we are able to understand that "*He*" in the second sentence is referring to Andy Jassy. Without this context of the subject in the first sentence, it is difficult to understand the relationship between the rest of the words in the text.

Now that we've reviewed transformers and explained their importance in NLP, how does this relate to CV? The vision transformer was introduced in a 2021 paper, *An Image is Worth 16x16 Words: Transformers for Image Recognition at Scale* (`https://arxiv.org/pdf/2010.11929v2.pdf`). Vision transformers expand upon the concept of text transformers. The technical details of vision transformers are outside of the scope of this book; however, they have shown great improvement over CNNs for image classification tasks. Transformer architecture has introduced new innovations such as generative AI. Generative AI is blurring the lines between generating separate models for NLP and CV. With generative AI, we can generate images from a text phrase. One image generator developed by OpenAI is called DALL-E (`https://openai.com/blog/dall-e-now-available-without-waitlist/`). Another example created by **Stability AI** is called **Stable Diffusion** (`https://huggingface.co/spaces/stabilityai/stable-diffusion`). All that is required to generate an image is to type in an English phrase. The following figure shows an example of images generated by **Stable Diffusion**:

Figure 1.8 - Images created from text "An astronaut in the mountains" using Stable Diffusion

The potential use cases for transformers and generative AI are just beginning to be explored. Throughout the rest of this book, we will discuss the following real-world applications of CV and provide code examples.

Contactless check-in and checkout

To improve the customer experience, many businesses have adopted contactless check-in and checkout processes. This provides a frictionless experience that reduces cost and is easier to scale. It also adds a layer of enhanced security. Instead of checking out from a grocery store using a credit card or trying to remember a PIN, you can use a biometric option such as your palm or facial recognition. In *Chapter 4*, we will walk through a code example to build a contactless hotel check-in system using identity verification.

Video analysis

You can use CV to analyze videos to detect objects in real time. This helps gather analytics for security footage and helps ensure compliance requirements are met in a manufacturing facility. In the media and entertainment industry, companies can monetize their content by automating the analysis of videos to determine when to insert advertisements. In *Chapter 5*, we will use CV to annotate and automate security video footage.

Content moderation

The amount of digital content is increasing. Often, this content is moderated manually by human reviewers, which is not a scalable or cost-effective solution. Companies in the gaming, social media, financial services, and healthcare industries are looking to protect their brands, create safe online communities that improve the user experience, meet regulatory and compliance requirements, and reduce the cost of content moderation. CV services combined with additional AI services, such as NLP, can automatically moderate image, video, text, and audio workflows to detect unwanted or offensive content and protect sensitive information. In *Chapter 6*, we teach you how to incorporate these capabilities into an automated pipeline.

CV at the edge

CV at the edge allows you to run your models locally on an edge device to make real-time predictions and reduce latency. Many ML use cases require that models run on the edge. To meet privacy preservation standards, users' data needs to be kept directly on devices such as mobile phones, smart cameras, and smart speakers. Also, your devices may be running in places with limited connectivity such as oil drills, or even in space where it is impossible to send your data to the cloud to perform inference. Consultancy firm Deloitte estimates that today there are over 750 million AI devices, and that number is only continuing to grow. What types of use cases can CV solve at the edge? Camera streams on a manufacturing floor can trigger multiple models and alert maintenance teams when equipment defects are identified and can also detect issues in product quality. It also has applications in healthcare. CV models can be deployed on X-ray machines and in operating rooms to quickly process medical images, which helps with faster patient diagnosis. In *Chapters 7* and *8*, we'll dive deeper into CV at the edge and provide code examples to solve industrial Internet of Things (IoT) scenarios and defect detection.

In this section, we introduced transformers and discussed their impact on CV. We also covered common challenges that can be solved with CV across multiple industries. These use cases are not an exhaustive list and represent only a small sample of how CV can unlock meaningful insights from your content and accelerate your business outcomes. In the next section, we will introduce the AWS AI/ML services and the benefits of using these services in your downstream applications.

Exploring AWS AI/ML services

There are many challenges faced when building and deploying a production CV model. It's often difficult to find the right ML skill sets. Gathering high-quality data and labeling the data is a manual and costly process. Data processing and feature engineering require domain expertise. Developing, training, and testing ML models takes time. Once a model is created and deployed into production, it's challenging to scale on-premises and difficult to understand which metrics to monitor to detect data and model quality drift. Reducing inference latency, automating the retraining process, and managing the underlying infrastructure are also concerns.

AWS AI/ML services are designed to address these challenges. These services are fully managed, so you don't have to worry about their underlying architecture. You can also optimize your costs by only paying for what you use. Within the portfolio of AWS AI/ML services, there are several approaches to choose from when building your CV application.

AWS AI services

AWS AI services provide pre-trained models that use DL technology to solve common use cases such as image classification, personalized recommendations, fraud detection, anomaly detection, and NLP. These services don't require any ML expertise and they're easily integrated into your applications or with other AWS services by calling APIs. They help remove the undifferentiated heavy lifting of dealing with image preprocessing and feature extraction. This way, you can focus on solving your business problems and moving to production faster.

One of the AI services for CV is **Amazon Rekognition**. It is a fully managed DL-based service that detects objects, people, activities, scenes, text, and inappropriate content in images and videos. It also provides facial analysis and facial search capabilities. Rekognition contains pre-trained models but also allows you to train your own custom model using **Rekognition Custom Models**. In the next two chapters, we provide code examples and applications of Rekognition and Rekognition Custom Models.

Amazon Lookout for Vision (`https://aws.amazon.com/lookout-for-vision/`) is another AI service that uses CV to detect anomalies and defects in manufacturing. Using pre-trained models, it helps improve industrial quality assurance by analyzing images to identify objects with visual defects. This helps improve your operational efficiency. In *Chapter 7*, we go into more detail about using Lookout for Vision.

For building and managing CV applications at the edge, **AWS Panorama** (`https://aws.amazon.com/panorama/`) provides ML devices and a **software development kit (SDK)** to add CV to your cameras.

This helps to automate costly inspection tasks by building CV applications to analyze video feeds. The Panorama appliance performs predictions locally for real-time decision-making. With Panorama, you can train your own models or select pre-built applications from AWS or third-party vendors.

These are only a few examples of the AWS AI services we will be focusing on in this book. For more details on the pre-trained services available for your applications, visit the AWS **Machine Learning | AI Services** page (https://aws.amazon.com/machine-learning/ai-services/).

Amazon SageMaker

If you are interested in fine-tuning a pre-trained model, using built-in algorithms, or building your own custom ML model, **Amazon SageMaker** (https://aws.amazon.com/sagemaker/) is a comprehensive fully managed ML service that allows you to prepare data, build, train, and deploy ML models for any use case. SageMaker provides the infrastructure, tools, visual interfaces, workflows, and MLOps capabilities for every step of the ML life cycle to help you deploy and manage models at scale. SageMaker also contains an **integrated development environment** (IDE) called SageMaker Studio where you can perform all steps within the ML life cycle and orchestrate **continuous integration/ continuous deployment** (CI/CD) pipelines. For more information on SageMaker Studio, refer to the book *Getting Started with SageMaker Studio*, by Michael Hsieh (https://www.packtpub.com/product/getting-started-with-amazon-sagemaker-studio/9781801070157):

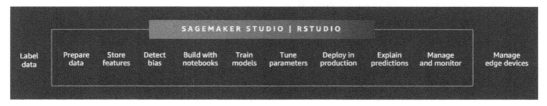

Figure 1.9 – Amazon SageMaker features and capabilities

With SageMaker, you can use **transfer learning** (TL) to fine-tune and reuse a pre-trained model without training a model from scratch. This saves you time and allows you to transfer the domain knowledge you gained previously to solve a new ML problem. This technique can be applied to CV or any type of business problem.

SageMaker contains dozens of pre-built algorithms that are optimized for speed, scale, and accuracy. They include support for supervised and unsupervised algorithms to solve a variety of use cases, including CV-related problems such as image classification, object detection, and semantic segmentation.

If a pre-trained or pre-built solution does not fit your needs, you have the option to build a custom ML model. There is a variety of powerful CPU and GPU compute options available for training and hosting your model on SageMaker. In *Chapter 12*, we will build a custom CV model on SageMaker to classify different types of skin cancer.

In this section, we provided an overview of the AWS AI/ML services related to CV. Next, we will show you how to set up the AWS environment that you will use throughout this book to build CV solutions.

Setting up your AWS environment

In the following chapters, you will need access to an AWS account to run the code examples. If you already have an AWS account, feel free to skip this section and move on to the next chapter.

> **Note**
> Please use the AWS Free Tier, which allows you to try services free of charge based on certain service usage limits or time limits. See `https://aws.amazon.com/free` for more details.

Follow the instructions at `https://docs.aws.amazon.com/accounts/latest/reference/manage-acct-creating.html` to sign up for an AWS account, then proceed as follows:

1. Once the AWS account is created, sign in using your email address and password and access the AWS Management Console at `https://console.aws.amazon.com/`.

2. Type `IAM` in the services search bar at the top of the console and select **IAM** to navigate to the IAM console. Select **Users** from the left panel in the IAM console and select on **Add User**.

3. Enter a **User name** value, then select **Programmatic access** and **AWS Management Console access** for **Access type**. Keep the **Console password** setting as **Autogenerated password**, and keep **Require password reset** as selected:

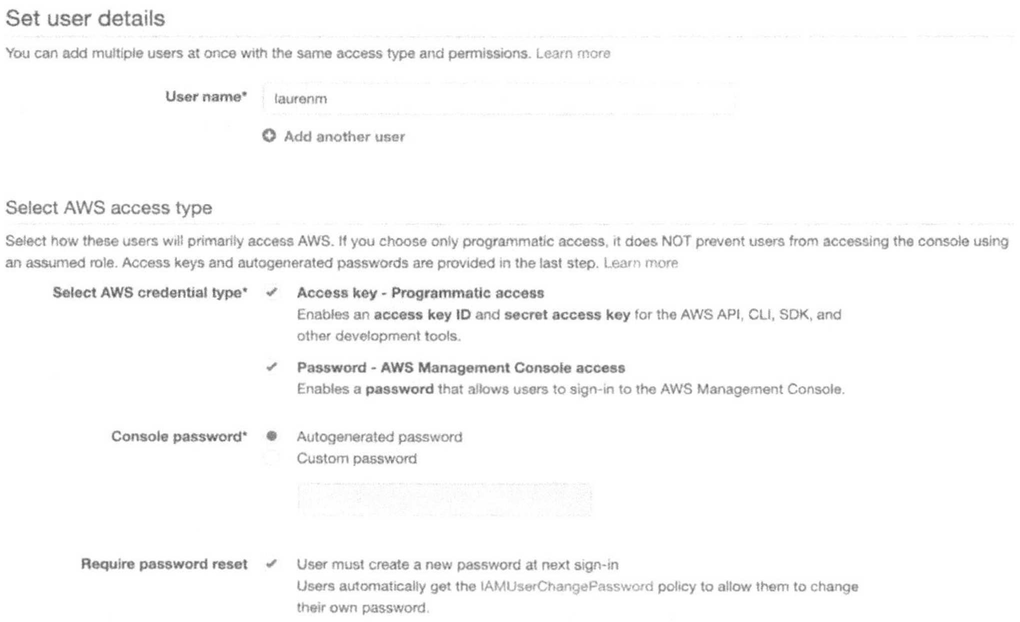

Set user details

You can add multiple users at once with the same access type and permissions. Learn more

User name*   laurenm

O Add another user

Select AWS access type

Select how these users will primarily access AWS. If you choose only programmatic access, it does NOT prevent users from accessing the console using an assumed role. Access keys and autogenerated passwords are provided in the last step. Learn more

Select AWS credential type*   ✓   Access key - Programmatic access
Enables an **access key ID** and **secret access key** for the AWS API, CLI, SDK, and other development tools.

✓   Password - AWS Management Console access
Enables a **password** that allows users to sign-in to the AWS Management Console.

Console password*   ● Autogenerated password
Custom password

Require password reset   ✓   User must create a new password at next sign-in
Users automatically get the IAMUserChangePassword policy to allow them to change their own password.

Figure 1.10 – Setting your IAM username and access type

4. Select **Next: Permissions**. On the **Set permissions** page, select on **Attach existing policies directly** and select the checkbox to the left of **AdministratorAccess**. Select **Next** twice to go to the **Review** page. Select **Create user**:

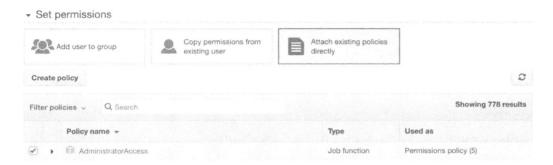

Figure 1.11 – Adding Administrator access for IAM user

5. Now, go back to the AWS Management Console (`console.aws.amazon.com`) and select **Sign In**. Provide the IAM username you created in the previous step along with a temporary password, and enter a new password to log in to the console.

Creating an Amazon SageMaker Jupyter notebook instance

We will be using Jupyter Notebooks to run our code in the following chapters. Please execute the following steps to create a notebook instance in Amazon SageMaker:

1. In the AWS Management Console, type `SageMaker` in the services search bar at the top of the page, and select on it to access the Amazon SageMaker console.

2. On the left panel, select on **Notebook** to expand, and select **Notebook instances**.

3. At the top right of the **Notebook instances** page, select **Create notebook instance**.

4. Under **Notebook instance settings**, type a name for **Notebook instance name**. For **Notebook instance type**, select **m1.t3.medium** since it falls under the AWS Free Tier:

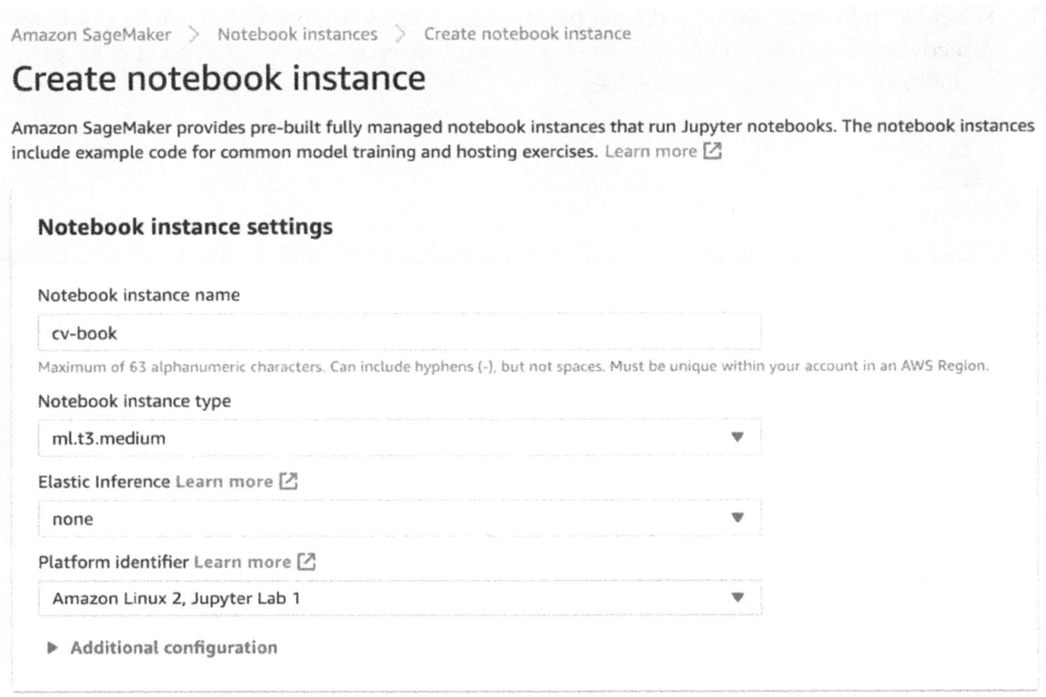

Figure 1.12 – Amazon SageMaker: Notebook instance settings

5. Under the **Permissions and encryption** section, select the **IAM role** list and choose **Create a new role**. Specify **Any S3 bucket** to provide access to all S3 buckets.

6. Leave the rest of the default options in the remaining sections and select **Create notebook instance**.

7. It will take a few minutes for the notebook instance to provision. Once the status is **InService**, you are ready to proceed. The following chapters will provide instructions for executing the code examples.

Now, you are ready to deploy the code examples that will show you how to use AWS AI/ML services to deploy CV solutions. Throughout the rest of the book, you will use a SageMaker notebook instance for these steps.

Summary

In this chapter, we covered the architecture behind a CV DNN and the common CV problem types. We discussed how to create high-quality datasets by preprocessing your input images, extracting features, and auto-labeling your data. Next, we summarized recent CV advancements and provided a brief overview of common CV use cases and their importance in deriving value for your business. We also explored AWS AI/ML services and how they can be used to quickly deploy production solutions.

In the next chapter, we will introduce Amazon Rekognition. You will learn about the different Rekognition APIs and how to interact with them. We will dive deeper into several use cases and provide Python code examples for execution.

2

Interacting with Amazon Rekognition

A picture is worth a thousand words because it captures so much context. Imagine looking at a recent photo or video of a family trip. Your two kids are playing on the beach with the dog, and it's sunny and warm. Their suits are still dry, so this must have occurred near your arrival time. There's an endless ocean of contextually sensitive details your eyes can detect.

Since computers don't have eyes, we need to use artificial intelligence to mimic this capability. Amazon Rekognition's mission is to make computer vision accessible to every developer. With Rekognition, you can build faster by leveraging its APIs to access high-quality results. The built-in APIs can detect the following within photos and images: objects, faces, scenes, activities, text, pathing, and segmentations. These capabilities simplify operations without requiring data science teams to build custom models, and you continuously receive improvements. Additionally, Rekognition is easily customizable, as you'll learn in the next chapter.

By the end of this chapter, you will have a better understanding of Amazon Rekognition. You will become familiar with how to do the following:

- Use the Amazon Rekognition Console
- Programmatically detect objects, activities, and scenes
- Use **Python Imaging Library** (**PIL**) to draw bounding boxes

Technical requirements

A Jupyter notebook is available for running the example code from this chapter. You can access the most recent code from this book's GitHub repository here: `https://github.com/PacktPublishing/Computer-Vision-on-AWS`.

You can clone that repository to your local machine using the following command:

```
$ git clone https://github.com/PacktPublishing/Computer-Vision-
on-AWS
$ cd Computer-Vision-on-AWS/02_IntroRekognition
```

Additionally, you will need an AWS account and Jupyter Notebook. *Chapter 1* contains detailed instructions for configuring the developer environment.

The Amazon Rekognition console

The AWS management console gives you secure access to your AWS resources. You can explore its functionality using any modern web browser, such as Google Chrome, Microsoft Edge, or Mozilla Firefox. Most customers access the management console at `https://console.aws.amazon.com/`. Confirm the endpoint with your IT support department if you use an employer's AWS account.

Within the web application is the Amazon Rekognition service console. It contains a series of image **Demos** and **Video Demos** pages. These **single-page applications** (**SPAs**) illustrate various core scenarios of the Amazon Rekognition service. You can also find links to product documentation, custom label design wizards, and troubleshooting metrics.

First, let's examine using the Amazon Rekognition console. To access it, take the following steps:

1. Navigate to `https://console.aws.amazon.com/rekognition/`.

2. Choose **Ohio** (**us-east-2**), or another supported Region, using the Regional selector in the top-right corner.

Figure 2.1 – Amazon Rekognition's management console

Using the Label detection demo

In the left-hand shortcut list, expand **Demos**, and choose **Label detection**. This demonstration page lets you interact with sample images or upload a custom image.

The default image contains a skateboarder kickflipping across the street. Amazon Rekognition detects and scores that this image includes a person, a skateboard, and multiple vehicles. The many object detections also include a bounding box representing an object's location in the photo.

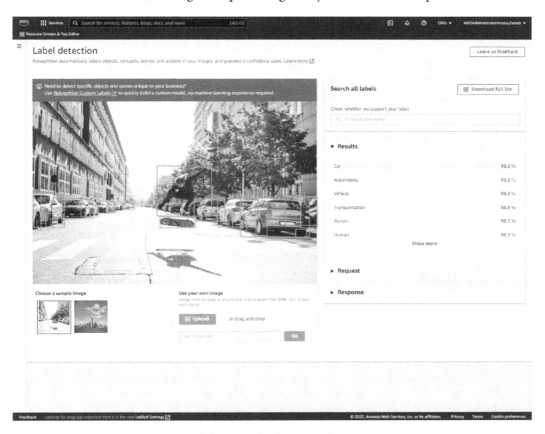

Figure 2.2 – Amazon Rekognition's Label detection demonstration page

The second sample image demonstrates scene and concept detection. Activate the city example to view label predictions like this is an urban or metropolis environment full of buildings. You'll notice that these labels don't contain bounding boxes because scene detection considers the entire image.

Next, use the **Upload** button to experiment with images from the `02_IntroRekognition` folder. You can also specify pictures from your photo library. Try mixing and matching images of objects, scenes, and actions. The demo pages maintain confidentiality and don't save your custom images.

> **Important note**
>
> Amazon Rekognition requires images to be in JPEG or PNG format and no larger than 5 MB.

Examining the API request

This demonstration page includes the payload used to call Amazon Rekognition's `DetectLables` API. Here, it specifies the image is available within an **Amazon Simple Storage Service** (**Amazon S3**) bucket.

In this case, Rekognition analyzes the `skateboard.jpg` file inside the `recognition-console-sample-images-prod` bucket. You will create and secure a custom bucket for your images in the proceeding section:

```
{
  "Image": {
    "S3Object": {
      "Bucket": "rekognition-console-sample-images-prod",
      "Name": "skateboard.jpg"
    }
  }
}
```

Examining the API response

Expand the Response accordion control to examine the API's output. We'll begin with the response's envelope, which consists of the `LabelModelVersion` and `Labels` properties (see *Table 2.1*).

The `LabelModelVersion` property increases very slowly because it refers to the broader feature. AWS continuously improves its detection models and doesn't report the revision component (e.g., 2.0.xxxx).

Suppose the input image contains **Exchangeable Image File Format** (**Exif**) metadata. In that case, you might observe that the envelope reports an `OrientationCorrection` value:

```
{
  "Labels": [
    ...
  ],
  "LabelModelVersion": "2.0"
}
```

The following table enumerates the `DetectLabels` response properties. This envelope structure wraps the `Labels` array and includes various additional metadata:

Property	Type	Description
`LabelModelVersion`	String	The version number of the label detection model.
`Labels`	An array of `Label` objects	Collection of real-world objects, scenes, and activities detected.
`OrientationCorrection`	String	Rotating the image 0, 90, 180, or 270 degrees has happened.

Table 2.1 – The DetectLabels response

3.   A `Label` is a structure containing details about the detected label, including the name, specific instances, parents, and confidence scores (see *Table 2.1*). You can find these structures within the response envelope's `Labels` property.

For instance, this example label denotes that Amazon Rekognition is 98.9% confident the image contains a car. Since this label refers to an object, you'll receive the specific *instances* within the image. Suppose the `Label` object refers to the scene, such as whether it's sunny or urban. In that case, `Instances` will be empty because Amazon Rekognition is reporting the whole image has this characteristic:

```
{
    "Name": "Car",
    "Confidence": 98.87621307373047,
    "Instances": [
        . . .
    ],
    "Parents": [
        . . .
    ]
},
```

The following table enumerates the `Label` object's properties. Suppose the `Label` object has a low `Confidence` score. In that case, you should consider it erroneous and ignore it until *Chapter 11*. This same recommendation applies to all objects that contain a `Confidence` property:

Property	Type	Description
Confidence	Float	Level of confidence.
Instances	An array of `Instance` objects	If `Label` represents an object, `Instances` contains the bounding boxes for each instance of the detected object. Bounding boxes are returned for standard object labels such as people, cars, furniture, apparel, or pets.
Name	String	The name of the label.
Parents	An array of `Parent` objects	The parent labels for this label.

Table 2.2 – Label properties

Amazon Rekognition found two instances of cars for this specific detection. An *instance* of a label has two properties, `BoundingBox` and the `Confidence` score (see *Table 2.3*). `BoundingBox` defines the normalized image coordinates containing the object. Later in this chapter, you'll learn how to denormalize these values by multiplying the image size:

```
"Instances": [
    {
        "BoundingBox": {
            "Width": 0.10527367144823074,
            "Height": 0.18472492694854736,
            "Left": 0.0042892382480204105,
            "Top": 0.5051581859588623
        },
        "Confidence": 98.87621307373047
    },
    {
        "BoundingBox": {
            "Width": 0.028528062626719475,
            "Height": 0.05612713471055031,
            "Left": 0.26153871417045593,
            "Top": 0.5507346987724304
```

```
    },
    "Confidence": 60.064884185791016
}
```

When `Label` represents an object, one or more `Instances` specify its `BoundingBox` (location) within the image:

Property	Type	Description
BoundingBox	The BoundingBox object	The position of the label instance within the image.
Confidence	Float	The confidence that Amazon Rekognition has in the accuracy of the bounding box position.

Table 2.3 – The Instance object's properties

You'll notice that the labels have `Parents`, for example, `Car` has `Transportation` and `Vehicle`. This design allows you to write automation for your use case at granular or coarse fidelity:

```
{
    "Name": "Car",
    "Confidence": ...,
    "Instances": [
      ...
    ],
    "Parents": [
      {
        "Name": "Vehicle"
      },
      {
        "Name": "Transportation"
      }
    ]
},
```

The following table enumerates the `Parent` object's properties. Each `Label` object has zero or more parents:

Property	Type	Description
Name	String	The name of the parent label.

Table 2.4 – The Parent object's properties

Now that you understand the entire structure, you can review the `response.json` file in this chapter's repository.

Other demos

The Amazon Rekognition console contains demonstration pages for other features such as **Image moderation** and **Facial analysis**. You will learn how to use these capabilities in later chapters.

Monitoring Amazon Rekognition

Selecting **Metrics** navigates to an Amazon CloudWatch Dashboard. You can visit this page to quickly gain insights into API rates, failures, and throttling:

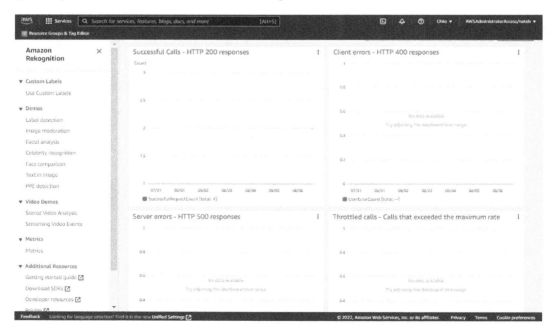

Figure 2.3 – Amazon Rekognition Metrics

The following table enumerates the built-in operational Amazon CloudWatch metrics for the `Rekognition` namespace. You specify metric thresholds with Amazon CloudWatch alerts to send emails, invoke custom code, and trigger third-party incident response software:

Metric	Description
SuccessfulRequestCount	The total number of successful requests.
ThrottledCount	The total number of throttled requests.
ResponseTime	The count of `Data Samples` or total milliseconds for Amazon Rekognition to compute the response.
ServerErrorCount	The total number of server errors.
UserErrorCount	The total number of invalid requests (e.g., missing parameters).

Table 2.5 – Built-in operational metrics

Important note
Amazon Rekognition's default service quotas might be too low for production environments. You can request quota increases by expanding the current user context in the top-right corner. Then, select **Service Quotas** from the drop-down menu.

Quick recap

In this section, you learned about the Amazon Rekognition console and how it quickly demonstrates the built-in APIs. This web app helps you learn about the service but isn't very customizable. To bring computer vision into our applications, we'll need to access Amazon Rekognition through its API.

Detecting Labels using the API

Developers interact with Amazon Rekognition using the **AWS Command-Line Interface** (**AWS CLI**), the AWS SDK, and REST clients. This section will use the `boto3` module for Python.

Uploading the images to S3

Use the AWS CLI or the Amazon S3 console to create a bucket for your test images. This example command will provision the cv-on-aws-book-**xxxx** bucket in **Ohio Region (us-east-2)**. Bucket names must be globally unique, so specify any random suffix. Next, record this value as you'll need it later:

```
$ aws s3 mb --region us-east-2 s3://cv-on-aws-book-nbachmei
make_bucket: cv-on-aws-book-nbachmei
```

Next, upload the sample files from the book's GitHub repository. You can complete this step using the following command:

```
$ aws s3 sync 02_IntroRekognition/images s3://cv-on-aws-book-
nbachmei/chapter_02/images --region us-east-2
```

Initializing the boto3 client

Open your Jupyter Notebook using the steps from *Chapter 1*. Import `DetectLabel.ipynb` from this chapter's GitHub repository. Then run the following snippet to create the Amazon Rekognition client.

Ensure the `bucket_name` and `region_name` variables equal the values from creating the bucket in the last step:

```
import boto3

bucket_name = 'cv-on-aws-book-nbachmei'
region_name = 'us-east-2'
recognition = boto3.client('rekognition', region_name=region_
name)
```

Running this snippet will initialize the Amazon Rekognition client for further use.

> **Important note**
> An error such as `ModuleNotFoundError: No module named 'boto3'` means you need to install the `boto3` module. In that case, run the `!python3 -m pip install boto3` command within a notebook code cell.

Detect the Labels

You use the `DetectLabel` API to analyze the `skateboard.jpg` image programmatically. This operation requires passing a payload such as the **Label detection** page:

```
response = rekognition.detect_labels(
    Image={
        'S3Object': {
            'Bucket': bucket_name,
            'Name': 'chapter_02/images/skateboard.jpg'
        }
    })
print(response)
```

By default, this API returns up the first 100 labels with a confidence score greater than or equal to 55 percent. Suppose you want to restrict the responses to those with 80 percent confidence or only the first 10 labels. In that case, specify the optional `MinConfidence` and `MaxLabels` parameters:

```
filtered_response = rekognition.detect_labels(
    Image={
        'S3Object': {
            'Bucket': bucket_name,
            'Name': 'chapter_02/images/skateboard.jpg'
        },
    },
    MinConfidence= 80,
    MaxLabels= 10
    )


print(filtered_response)
```

Several Amazon Rekognition APIs, such as `DetectLabels`, support input images as base64-encoded objects. This option benefits ephemeral photos or when you'll only keep a small subset of the analyzed content:

```
with open('images/skateboard.jpg','rb') as f:
    image_bytes = f.read()

    labels = rekognition.detect_labels(
        Image={
            "Bytes": image_bytes
        })
    print(labels)
```

Using the Label information

The `DetectLabels` API returns a JSON document containing a list of labels. Those labels form a hierarchical structure with zero or more parent labels. Let's render the label tree using the `treelib` module. While more powerful modules exist, such as `Graphziv` and `PyDot`, `treelib` focuses on simplicity:

```
$ python3 -m pip3 install treelib
```

This module has two classes: Tree and Node. The tree represents a container for zero or more nodes, and each node represents a hierarchical tree element. For instance, you could describe the previous section's Car label using this snippet:

```
from treelib import Tree, Node

myTree = Tree()
transportation_node = myTree.create_node('Transportation')
vehicle_node = myTree.create_node('Vehicle',
parent=transportation_node.identifier)
car_node = myTree.create_node('Car', parent=vehicle_node.
identifier)
sedan_node = myTree.create_node('Sedan', parent=car_node.
identifier)

myTree.show()
```

Running this code in your Jupyter notebook will output a simple tree like the following:

```
Transportation
└── Vehicle
    └── Car
        └── Sedan
```

Manually building the tree helps understand the module, but you'll typically need to create these representations dynamically. Let's start by parsing the Detect Label's response envelope into the tree.

The first step is starting another tree using the following snippet. It defines another_tree and appends the root node called Response. Next, the code adds child elements for LabelModelVersion and Labels:

```
from treelib import Tree, Node

another_tree = Tree()
root = another_tree.create_node('Response')

model_version_node = another_tree.create_node(
    'LabelModelVersion : %s' % response['LabelModelVersion'],
    parent=root.identifier)
```

```
labels_node = another_tree.create_node(
    'Labels: %d total' % len(response['Labels']),
    parent=root.identifier)

another_tree.show()
```

Running this snippet will return output like the following example:

```
Response
├── LabelModelVersion : 2.0
└── Labels: 29
```

Next, let's traverse the response document and build child nodes for the top 10 highest Confidence labels. Use the built-in sorted function keyed on label[' Confidence'] in reverse descending order. Then, use the array slice operator to limit the results. Finally, iterate through the objects to build the tree:

```
sorted_by_confidence = sorted(
    response['Labels'],
    key=lambda label: label['Confidence'],
    reverse=True)

top_10 = sorted_by_confidence[:10]

for label in top_10:
    name = label['Name']
    confidence = label['Confidence']
    label_node = another_tree.create_node(
        '%15s - [Confidence: %2.2f%%]' % (name, confidence),
        parent=labels_node.identifier)
```

When you call the another_tree.show function, it should output results like the following:

> **Important note**
> The treelib module doesn't guarantee a specific child node order at render time. Your specific output might vary, such as Vehicle reports before Machine.

```
Response
├── LabelModelVersion : 2.0
└── Labels: 29 total
    ├──              Car - [Confidence: 99.15%]
    ├──             Road - [Confidence: 92.82%]
    ├──            Human - [Confidence: 98.99%]
    ├──            Wheel - [Confidence: 93.25%]
    ├──           Person - [Confidence: 98.99%]
    ├──          Machine - [Confidence: 93.25%]
    ├──          Vehicle - [Confidence: 99.15%]
    ├──       Automobile - [Confidence: 99.15%]
    ├──       Pedestrian - [Confidence: 92.78%]
    └──   Transportation - [Confidence: 99.15%]
```

Lastly, you can extend this tree with child nodes representing individual label instances and parent names. Not every label has parents or even instances; in those situations, the Detect Labels API returns an empty set:

```
instances = sorted(
    label['Instances'],
    key=lambda instance: instance['Confidence'],
    reverse=True)


if len(instances) > 0:
    instances_node = another_tree.create_node(
        'Instances: %d' % len(instances),
        parent=label_node.identifier)


    another_tree.create_node(
        'Max Confidence: %2.2f%%' % instances[0]
['Confidence'],
        parent=instances_node.identifier)


    another_tree.create_node(
        'Min Confidence: %2.2f%%' % instances[-1]
['Confidence'],
        parent=instances_node.identifier)
```

```
parents = label['Parents']
if len(parents) > 0:
    for parent in parents:
        another_tree.create_node(
            'Parent: %s' % parent['Name'],
            parent=label_node)
```

Running this snippet within the Jupyter notebook will return a response, such as the following example output:

```
Response
├── LabelModelVersion : 2.0
└── Labels: 29 total
    ├── Automobile - [Confidence: 99.15%]
    │   ├── Parent: Transportation
    │   └── Parent: Vehicle
    ├── Car - [Confidence: 99.15%]
    │   ├── Instances: 14
    │   │   ├── Max Confidence: 99.15%
    │   │   └── Min Confidence: 52.38%
    │   ├── Parent: Transportation
    │   └── Parent: Vehicle
    ├── Human - [Confidence: 98.99%]
    ├── Machine - [Confidence: 93.25%]
    ├── Pedestrian - [Confidence: 92.78%]
    │   └── Parent: Person
    ├── Person - [Confidence: 98.99%]
    │   └── Instances: 2
    │       ├── Max Confidence: 98.99%
    │       └── Min Confidence: 85.03%
    ├── Road - [Confidence: 92.82%]
    ├── Transportation - [Confidence: 99.15%]
    ├── Vehicle - [Confidence: 99.15%]
    │   └── Parent: Transportation
    └── Wheel - [Confidence: 93.25%]
```

Finally, let's combine all code snippets and review the complete example:

```python
from treelib import Tree, Node

another_tree = Tree()
root = another_tree.create_node('Response')
model_version_node = another_tree.create_node(
    'LabelModelVersion : %s' % response['LabelModelVersion'],
    parent=root.identifier)
labels_node = another_tree.create_node(
    'Labels: %d total' % len(response['Labels']),
    parent=root.identifier)

sorted_by_confidence = sorted(
    response['Labels'],
    key=lambda label: label['Confidence'],
    reverse=True)

top_10 = sorted_by_confidence[:10]

for label in top_10:
    name = label['Name']
    confidence = label['Confidence']
    label_node = another_tree.create_node(
        '%s - [Confidence: %2.2f%%]' % (name, confidence),
        parent=labels_node.identifier)

    instances = sorted(
        label['Instances'],
        key=lambda instance: instance['Confidence'],
        reverse=True)

    if len(instances) > 0:
        instances_node = another_tree.create_node(
            'Instances: %d' % len(instances),
            parent=label_node.identifier)
```

```
        another_tree.create_node(
            'Max Confidence: %2.2f%%' % instances[0]
['Confidence'],
            parent=instances_node.identifier)

        another_tree.create_node(
            'Min Confidence: %2.2f%%' % instances[-1]
['Confidence'],
            parent=instances_node.identifier)

    parents = label['Parents']
    if len(parents) > 0:
        for parent in parents:
            another_tree.create_node(
                'Parent: %s' % parent['Name'],
                parent=label_node)

another_tree.show()
```

Using bounding boxes

For many scenarios, you'll want a bounding box that contains a detected object. These situations include improving image cropping, adding explainability and transparency, and troubleshooting, among others. Amazon Rekognition reports the bounding box for the Label instances that support this feature.

Suppose you want to draw a red box around the skateboard. In that case, you must find the label instance for Skateboard:

```
from json import dumps

def find_first_label(response, label_name):
    for label in response['Labels']:
        if label['Name'] == label_name:
            return label

    print('Unable to find a %s label' % label_name)
    return None

skateboard = find_first_label(response, 'Skateboard')
```

```
print(dumps(skateboard, indent=2))
```

Use the `dumps` function to print the skateboard, which should look like the following output:

```
{
  "Name": "Skateboard",
  "Confidence": 92.37877655029297,
  "Instances": [
    {
      "BoundingBox": {
        "Width": 0.12326609343290329,
        "Height": 0.058117982000112534,
        "Left": 0.44815489649772644,
        "Top": 0.6332163214683533
      },
      "Confidence": 92.37877655029297
    }
  ],
  "Parents": [
    {
      "Name": "Sport"
    },
    {
      "Name": "Person"
    }
  ]
}
```

Examining the `BoundingBox` properties reveals that every value is between zero and one. That's because the values are normalized, and you must multiply them by the image's dimensions to convert them back into pixel offsets. To achieve this, you need to do the following steps:

1. Download the image from the Amazon S3 bucket.

2. Use PIL to convert the raw bytes into an `Image` object.

3. Multiply `BoundingBox` by the `Image` dimensions.

First, let's initialize the Amazon S3 client and download the file:

```
import boto3

s3_client= boto3.client('s3', region_name=region_name)
image_file = s3_client.get_object(
    Bucket= bucket_name,
    Key= 'chapter_02/images/skateboard.jpg')


image_bytes = image_file['Body'].read()
```

You can install it using Python's `pip` model if this is your first time using PIL:

```
$ python3 -m pip install pillow
```

You can convert `image_bytes` into an image using the following snippet. It should report that the `skateboard.jpg` image has 1200 by 800 pixels width:

```
from PIL import Image
from io import BytesIO

image = Image.open(BytesIO(image_bytes))
print(image.size)
```

Then multiply the bounding box by these dimensions to denormalize the values. Finally, you can use the `ImageDraw` class to place a red rectangle around the skateboard:

```
from PIL import Image, ImageDraw

drawing = ImageDraw.Draw(image)
for instance in skateboard['Instances']:
    bounding_box = instance['BoundingBox']

    width = int(bounding_box['Width'] * image.size[0])
    left = int(bounding_box['Left'] * image.size[0])

    height = int(bounding_box['Height'] * image.size[1])
    top = int(bounding_box['Top'] * image.size[1])

    drawing.rectangle(
```

```
        xy=(left,top,left+width, left+height),
        outline='red')

image.show()
```

Figure 2.4 – Rendered Bounding Box

Quick recap

In this section, you learned how to use the `DetectLabels` API to analyze an image. You can implement similar code patterns to interact with more detection APIs, such as `DetectFaces` and `DetectProtectiveEquipment`. This table enumerates some of those operations:

API Name	Description
DetectCustomLabels	Finds instances of custom labels within the image.
DetectFaces	Finds instances of faces within the image.
DetectLabels	Finds predefined labels within the image.
DetectModerationLabels	Finds unsafe and inappropriate content within the image.
DetectProtectiveEquipment	Finds **personal protective equipment (PPE)** within the image.
DetectText	Finds text within the image.

Table 2.6 – Detection APIs

Cleanup

AWS will continue billing your account for any resources in use. In this chapter, you persist images into an Amazon S3 bucket and use Amazon Rekognition's `DetectLabels` API.

Assuming your account supports the free tier, you won't need to pay for the first 5 GB/month. Additionally, the first 5,000 `DetectLabels` API calls are also complementary.

Beyond that, Amazon S3 is around $0.023/GB per month for many regions. To clear all objects from your test bucket, review the *Emptying a bucket* documentation at `https://docs.aws.amazon.com/AmazonS3/latest/userguide/empty-bucket.html`. The `DetectLabels` API uses consumption-based pricing and will not receive ongoing charges from this lab.

Summary

When humans view visual information, we instinctively detect labels for objects, scenes, and activities. Amazon Rekognition offers similar capabilities as easy-to-use APIs that don't require machine learning expertise. Using the Rekognition management console, you learned how to upload messages and view the response payloads. Next, you programmatically repeated that process using the Python `boto3` module. Additionally, this chapter introduced the PIL for drawing bounding boxes on the example images.

Amazon Rekognition's built-in label detection supports over 2,500 labels with more than 250 pieces of supporting bounding box information. While this breadth covers numerous use cases, it won't cover every business-specific need, such as finding Packt Publishing's logo. In that case, you'll need to train Amazon Rekognition Custom Labels. Join us in the next chapter to learn how to use this feature without requiring undifferentiated heavy lifting or a data science team.

3
Creating Custom Models with Amazon Rekognition Custom Labels

In the last chapter, we learned about Amazon Rekognition's capability to identify objects in images and videos using built-in APIs. However, there are situations where you're looking for certain things that have specific meaning to you or your business. For example, imagine you're an auto parts manufacturer and want to identify different parts such as a crankshaft, torque converter, or radiator. Typically, generic **machine learning** (**ML**) models would identify these parts as **auto parts**, but that label may not be specific enough for your needs. Take another scenario—if you're a car enthusiast, you'd know the logos of different car manufacturers. If you would like to detect the logos of a car manufacturer using **computer vision** (**CV**), you'd need to build a customized ML model that can differentiate the logos of different car manufacturers. This is where Rekognition Custom Labels will come in handy. Custom Labels is a fully managed CV service that allows you to build models to classify and identify objects in images that are specific and unique to your business, with no ML experience required.

With Amazon Rekognition Custom Labels, you can identify objects and scenes in images that are specific to you or your business needs. You can use it to perform image classification (image-level predictions) or object detection (bounding box-level predictions). For example, if you are a grapevine farmer, you can detect the current growth cycle of grapes (flowering, veraison, or harvest-ready) using Amazon Custom Labels.

By the end of this chapter, you should have a better understanding of Amazon Rekognition Custom Labels and the process to train and deploy models to detect objects such as Packt's logo. You can similarly create and host models using Custom Labels to detect objects specific to your use case.

This chapter covers the following topics:

- Introducing Amazon Rekognition Custom Labels
- Creating a model using Rekognition Custom Labels

- Building a model to identify Packt's logo
- Validating that the model works

Technical requirements

You will require the following:

- Access to an active **Amazon Web Services** (**AWS**) account with permissions to access Amazon SageMaker and Amazon Rekognition
- PyCharm or any Python IDE
- All the code examples for this chapter can be found on GitHub at `https://github.com/PacktPublishing/Computer-Vision-on-AWS`

A Jupyter notebook is available for running the example code from this chapter. You can access the most recent code from this book's GitHub repository, `https://github.com/PacktPublishing/Computer-Vision-on-AWS`. Clone that repository to your local machine using the following command:

```
$ git clone https://github.com/PacktPublishing/Computer-Vision-
on-AWS
$ cd Computer-Vision-on-AWS/03_RekognitionCustomLabels
```

Additionally, you will need an AWS account and Jupyter notebook. *Chapter 1* contains detailed instructions for configuring the developer environment.

Introducing Amazon Rekognition Custom Labels

Developing a custom ML model to analyze images is a significant undertaking that requires tremendous time, ML expertise, and resources. Additionally, it generally requires thousands of hand-labeled images to provide the model with enough data to accurately make decisions. It would take months to gather this data and typically requires large teams of human labelers to prepare it for use in ML.

With Amazon Rekognition Custom Labels, you can offload this heavy lifting to the service. Custom Labels builds off of Amazon Rekognition's existing capabilities (as explained in *Chapter 2*), using **transfer learning** (**TL**). Instead of you needing to provide thousands of images, you can take a small set of images (typically around 100-200 images) for each label to train a model. If your images are already labeled, you can directly import them into Custom Labels. If not, you can use Custom Labels' built-in labeling interface or use SageMaker GroundTruth to label. Once you start training, Rekognition Custom Labels will inspect the images and select appropriate ML algorithms to train a model. This entire process is transparent to you as a user, and you will receive model performance metrics at the end of training.

Rekognition Custom Labels doesn't require you to have any prior CV expertise. You can get started by simply uploading tens of images instead of thousands. Using TL, Rekognition Custom Labels automatically inspects the training data, selects the right model framework and algorithm, optimizes the hyperparameters, and trains the model. When you're satisfied with the model's accuracy, you can start hosting the trained model with a single click.

Benefits of Amazon Rekognition Custom Labels

The following are some of the benefits of Amazon Rekognition Custom Labels:

- **Efficient data labeling**—The Amazon Rekognition Custom Labels console provides a visual interface to perform labeling on your images fast and simply. The interface allows you to apply a label to the entire image. Additionally, you can identify and label specific objects in images using bounding boxes with a click-and-drag interface. If you have a large dataset, you can use Amazon SageMaker Ground Truth to efficiently label your images at scale.

- **Automated ML (AutoML)**—You do not need ML expertise to build your model. Amazon Rekognition Custom Labels provides AutoML capabilities that take care of the ML for you. Once you provide the training images, Amazon Rekognition Custom Labels can automatically load and inspect the data, select the right ML algorithms, train a model, and provide model performance metrics.

- **Simplified model evaluation and inference**—With Amazon Rekognition Custom Labels, you evaluate your model's performance on your test dataset. For every image in the test dataset, you can compare the model's prediction versus the label assigned. You can also review detailed performance metrics such as precision, recall, F1 scores, and confidence scores. If you are satisfied with the model performance, you can start using your model immediately for image analysis. Alternatively, you can iterate and retrain new versions with more images to improve performance.

Now that we have gone through the basics of Rekognition Custom Labels, we will learn how to use them to create models for our use cases.

Creating a model using Rekognition Custom Labels

In this section, we will learn how to select the type of model you need for your use case, how to create training and test datasets, how to start model training, and how to analyze images using the trained model.

Deciding the model type based on your business goal

First, you should decide which type of model you want to train depending on your business goals. For example, you could train a model to find your logo in social media posts, identify your products on store shelves, or classify machine parts in an assembly line.

You can train the following types of models with Amazon Rekognition Custom Labels:

- **Object classification: to find objects, scenes, and concepts**—With this type of model, you predict objects, scenes, or concepts associated and represented in the entire image. It is also referred to as image classification. For example, you can train a model that identifies the type of fruit in an image, such as an apple, a banana, or grapes. You can also train a model that classifies images into multiple categories. For example, the previous image might have categories such as unripe, ripe, or rotten.

- **Object detection: to find object locations**—With this type of model, you predict the location of an object on an image. The prediction includes bounding-box information for the object's location and a label that identifies the object within the bounding box. For example, you can train a model that shows bounding boxes around a rooftop and solar panels located on it.

Creating a model

Once you have decided on the type of model you want to train, you can start creating it. To create a model with Amazon Rekognition Custom Labels, you will need to create a project and training and test datasets, and initiate the training process for the model.

Creating a project

A project with Amazon Rekognition Custom Labels is a group of resources needed to create and manage a model. A project contains and manages the following:

- **Datasets**—A project has training and test datasets (images) and image labels that are used to train a model.

- **Models**—A project contains ML models you train to find scenes, objects, and their locations. You can have more than one version of a model in a project.

A project typically is used for a single use case, such as classifying LEGO bricks, distinguishing ripe and rotten fruits, or identifying car manufacturer logos.

Creating training and test datasets

A dataset is a set of images and labels that describe those images. Within your project, you create a training dataset and a test dataset that Amazon Rekognition Custom Labels uses to train and test your model.

A label identifies an object, scene, concept, or bounding box around an object in an image. Labels are either assigned to an entire image (image-level) or they are assigned to a bounding box that surrounds an object on an image.

You can create training and test datasets using the Rekognition console or the AWS SDK.

Creating training and test datasets using the AWS Management Console

You can start a project with a single dataset or separate training and test datasets. If you start with a single dataset, Amazon Rekognition Custom Labels splits your dataset during training to create a training dataset (80%) and a test dataset (20%) for your project. Start with a single dataset if you want Amazon Rekognition Custom Labels to decide which images are used for training and testing. For complete control over training, testing, and performance tuning, you should start your project with separate training and test datasets.

To create the datasets for a project, you import the images in one of the following ways:

- Import images from your local computer.

- Import images from a **Simple Storage Service** (S3) bucket. Amazon Rekognition Custom Labels can label the images using the names of the folders that contain the images.

- Import an Amazon SageMaker Ground Truth manifest file.

- Copy an existing Amazon Rekognition Custom Labels dataset.

Depending on where you import your images from, your images might be unlabeled. For example, images imported from a local computer aren't labeled, but images imported from an Amazon SageMaker Ground Truth manifest file are labeled. You can use the Amazon Rekognition Custom Labels console to add, change, and assign labels.

Initiating training for the model

Once you have created training and test datasets, the next step is to train the model. A new version of a model is created each time it is trained. During training, Rekognition Custom Labels tests the performance of your trained model. You can use the results to evaluate and improve your model. Training takes a while to complete. You are only charged for successful model training. If model training fails, Amazon Rekognition Custom Labels provides debugging information that you can use.

Improving the model

Once the training completes, you can evaluate the model using the metrics provided by Rekognition Custom Labels. Performance metrics such as F1, precision, and recall allow you to understand the performance of your model and help you decide if you're ready to use it in production. If performance improvements are needed, you can add more training images or improve dataset labeling.

> **Note**
>
> Precision is the fraction of correct predictions (also known as true positives) over all predictions (true positives + false positives) for an individual label. A higher value suggests higher precision (and better model performance).
>
> Recall is the fraction of label occurrences that were successfully predicted correctly for an individual label. A higher value suggests higher recall (and better model performance).
>
> The F1-score metric is a measure of the model performance of each label; a high value for the F1 score indicates the model is performing well for both precision and recall.

Starting your model

Once you're happy with the performance of the model, you need to start the model before you can use it. By starting the model, Rekognition will deploy and host your model on compute resources. You can start the model by using the Amazon Rekognition Custom Labels console or the `StartProjectVersion` API. You will be charged for the amount of time that your model runs.

Analyzing an image

Once the model is hosted, you can start analyzing your images using the `DetectCustomLabels` API. As part of the API call, you will supply either a local image or an image stored in an S3 bucket, along with the **Amazon Resource Name** (**ARN**) of the model that you want to use.

Stopping your model

As mentioned before, you are charged for the time that your model is running, regardless of if you're actively using it or not. If you are no longer using your model, stop the model by using the Rekognition Custom Labels console or by using the `StopProjectVersion` API.

Now that we have gone through the process of creating Rekognition Custom Labels, let's learn how to train and evaluate a model using Rekognition Custom Labels to identify Packt's logo.

Building a model to identify Packt's logo

In this section, we will see how to train a model to identify Packt's logo using Rekognition Custom Labels. We'll collect training and test datasets for Packt's logo, label the images and draw bounding boxes, and train the model.

Step 1 – Collecting your images

Upload the sample images from the book's GitHub repository. You can complete this step using the following command:

```
$ aws s3 sync 03_RekognitionCustomLabels/images s3://cv-on-aws-
book-xxxx/chapter_03/images --region us-east-2
```

> **Important note**
>
> To collect the sample images, you can use the same S3 bucket you created in *Chapter 2*.

Now, navigate to **Amazon Rekognition** on the AWS Management Console (https://us-east-2.console.aws.amazon.com/rekognition/home?region=us-east-2#/). Select on **Use Custom Labels** on the left sidebar and then Select on **Get started**:

Figure 3.1: Amazon Rekognition console

> **Important note**
>
> If you're a first-time user of the service, it will ask permission to create an S3 bucket to store your project files. Select **Create S3 bucket**.

Step 2 – Creating a project

Next, navigate to the **Projects** panel on the left sidebar and select on **Create project**:

Figure 3.2: Amazon Rekognition Custom Labels console

Give it some name, such as `Packt-logo-detection`.

Step 3 – Creating training and test datasets

Next, as we saw in the prior section of creating a model, we need to create a dataset. Select on **Create dataset**:

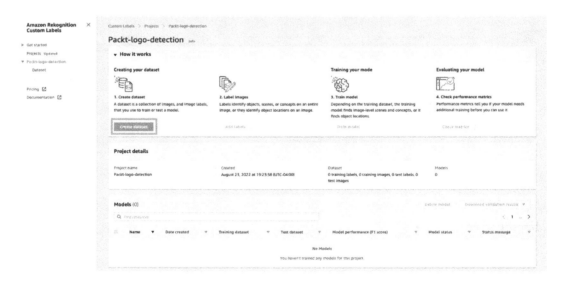

Figure 3.3: Managing projects on the Rekognition Custom Labels console

On the **Create dataset** page, leave the default configuration option of starting with a single dataset (where the service will split the dataset into training and test datasets):

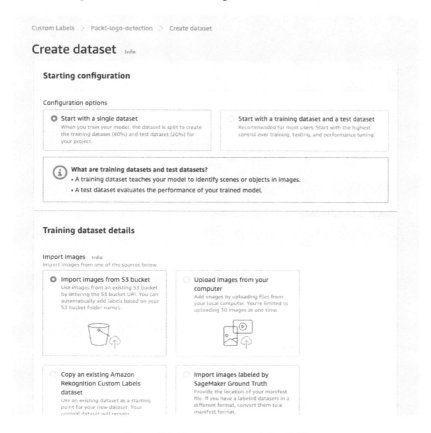

Figure 3.4: Creating a new dataset (1)

Under **Training dataset details**, select **Import images from S3 bucket** and provide the S3 URI for your bucket where you copied the images in *step 1*. The S3 URI would look similar to this: `s3://cv-on-aws-book-xxxx/chapter_03/images/`.

> **Important note**
> Keep in mind to add the trailing / in the S3 URI.

We'll leave the **Automatic labeling** box unchecked for this project, but you can use the feature to have Rekognition automatically assign image-level labels to images based on the folder where your images are stored:

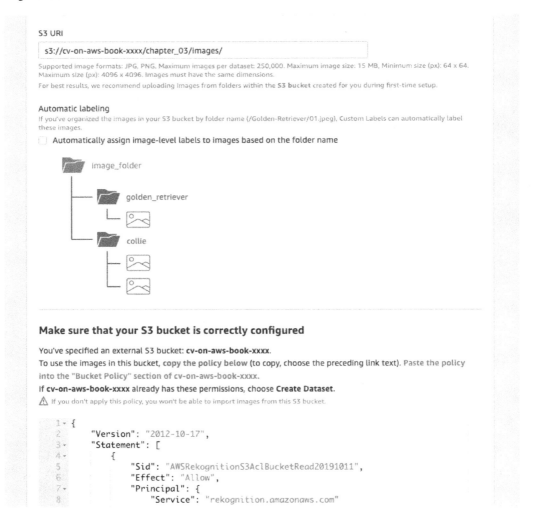

Figure 3.5: Creating a new dataset (2)

The last step in creating a dataset is to make sure your S3 bucket is configured correctly to allow Rekognition to access the images:

```
14          },
                    "Resource": "arn:aws:s3:::cv-on-aws-book-xxxx"
15          },
16          {
17              "Sid": "AWSRekognitionS3GetBucket20191011",
18              "Effect": "Allow",
19              "Principal": {
20                  "Service": "rekognition.amazonaws.com"
21              },
22              "Action": [
23                  "s3:GetObject",
24                  "s3:GetObjectAcl",
25                  "s3:GetObjectVersion",
26                  "s3:GetObjectTagging"
27              ],
28              "Resource": "arn:aws:s3:::cv-on-aws-book-xxxx/*"
29          },
30          {
31              "Sid": "AWSRekognitionS3ACLBucketWrite20191011",
32              "Effect": "Allow",
33              "Principal": {
34                  "Service": "rekognition.amazonaws.com"
35              },
36              "Action": "s3:GetBucketAcl",
37              "Resource": "arn:aws:s3:::cv-on-aws-book-xxxx"
38          },
39          {
40              "Sid": "AWSRekognitionS3PutObject20191011",
41              "Effect": "Allow",
42              "Principal": {
43                  "Service": "rekognition.amazonaws.com"
44              },
45              "Action": "s3:PutObject",
46              "Resource": "arn:aws:s3:::cv-on-aws-book-xxxx/*",
47              "Condition": {
48                  "StringEquals": {
49                      "s3:x-amz-acl": "bucket-owner-full-control"
50                  }
51              }
52          }
53      ]
54 }
```

Cancel   Create Dataset

Figure 3.6: Creating a new dataset (3)

Important note
If you don't apply the bucket policy to your S3 bucket, you won't be able to import images and won't be able to progress further.

Step 4 – Adding labels to the project

Once the images are imported into the Rekognition project, we need to add labels for each type of object, scene, or concept in your dataset. For our dataset, we will create a single label: `packt`.

To add the label, select on **Start labeling**:

Figure 3.7: Labeling the dataset

Once you're in labeling mode, the **Add labels** button on the left bar next to **Labels** will be enabled. Select on **Add labels**:

Figure 3.8: Creating a new label

You can add labels manually or import a list of labels from another dataset in Rekognition Custom Labels.

For our project, we will add labels manually. Type `packt` in the **New label name** field and select on **Add label** next to it. Select **Save**.

Step 5 – Drawing bounding boxes on your training and test datasets

The next step in the process is to draw bounding boxes on Packt's logo in the images. As we're already in labeling mode, we can select images and select on **Draw bounding boxes**:

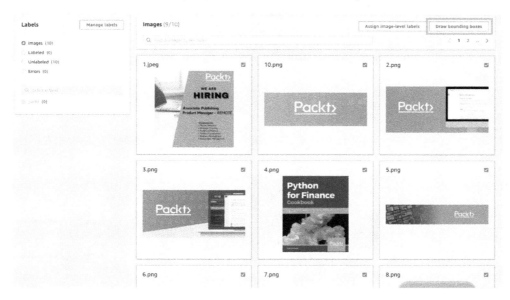

Figure 3.9: Managing the dataset

Once you select on **Draw bounding boxes**, you will be taken to a window where you can label the image/object and draw any bounding boxes. You can hover over the image and start drawing a box around Packt's logo:

Figure 3.10: Drawing a bounding box

Once you're done with labeling the image, you can select on the **Next** button on the top right-hand side (or press *Shift + Right arrow*).

> **Important note**
> You can use the additional capabilities of the labeling panel below the image to adjust or correct your bounding boxes or labels.

Finish labeling all the images in your dataset. Once you've completed labeling, you can select on **Save changes**:

Figure 3.11: Saving changes to the dataset

We'll now jump into the main part!

Step 6 – Training your model

Once we complete labeling all the images, the next step is to train a model using this dataset. As we discussed, Rekognition Custom Labels takes care of all the heavy lifting from a model training perspective, so all we need to do is select on **Train Model**:

Figure 3.12: Training details

> **Important note**
>
> You can add tags if you'd like to track your models. By default, Rekognition Custom Labels encrypt your data with a key that AWS owns and manages for you. However, you can also encrypt your images with your own encryption key through **AWS Key Management Service (AWS KMS)**.

As we haven't provided separate datasets for training and testing, Rekognition Custom Labels will split the dataset into a training dataset (80%) and a test dataset (20%). Select on **Train Model**.

The model training will take typically from 30 minutes to 24 hours to complete. You will be charged for the amount of time it takes to successfully train your model.

Validating that the model works

Once the model training finishes, you will see the performance of the trained model in the **Evaluation** tab. As you can see in the screenshot here, our trained model has really good precision and recall and a good F1 score for the `packt` label (with just eight training images):

Figure 3.13: Evaluating model performance

If you'd like to review how the model performed on your test dataset, select on **View test results**.

Step 1 – Starting your model

The next step is to start the model so that we can start using it to detect Packt's logo in images. You can go to the **Use model** tab and select on **Start**:

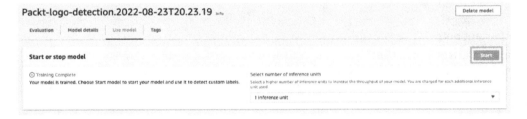

Figure 3.14: Starting the model using the Rekognition Custom Labels console

You will also need to select the number of inference units when you start your model— you will be charged for each inference unit.

> **Important note**
> A higher number of inference units will increase the throughput of your model, allowing you to process a greater number of images per second.

Step 2 – Analyzing an image with your model

Once the model has started, you can use it via the AWS CLI or the AWS SDK (such as Python Boto3). To make it easier to use the model, the Rekognition Custom Labels console provides sample code examples for CLI and Python scripts:

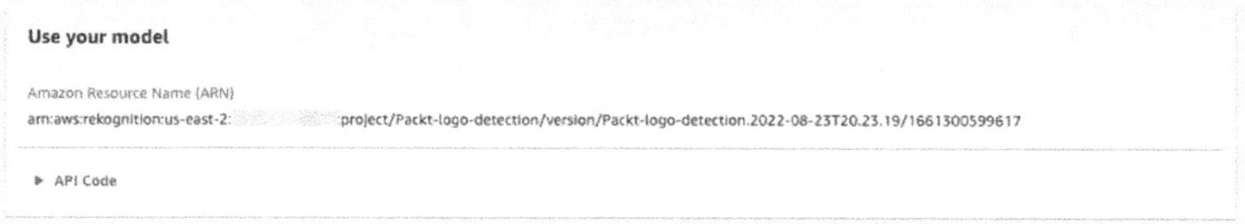

Use your model

Amazon Resource Name (ARN)
arn:aws:rekognition:us-east-2:   :project/Packt-logo-detection/version/Packt-logo-detection.2022-08-23T20.23.19/1661300599617

▶ API Code

Figure 3.15: Accessing sample code to use the trained model

In this case, Rekognition Custom Labels is analyzing the `locate_packt_logo.jpeg` file in your S3 bucket using the model you just started and returning detected labels with at least 85% confidence:

```
model='arn:aws:rekognition:us-east-2:xxxxxxxxxxxx:project/
Packt-logo-detection/version/Packt-logo-detection.2022-08-
23T20.23.19/1661300599617'
min_confidence=85/
response=rekognition.detect_custom_labels(
    Image={
        'S3Object': {
            'Bucket': bucket_name,
            'Name': 'chapter_03/locate_packt_logo.jpeg'
        }
    },
    MinConfidence=min_confidence,
    ProjectVersionArn=model
)
print(response)
```

The response of the API would look like the following:

```
{
    "CustomLabels": [
        {
            "Name": "packt",
            "Confidence": 91.34700012207031,
```

```
            "Geometry": {
                "BoundingBox": {
                    "Width": 0.08715999871492386,
                    "Height": 0.1518000066280365,
                    "Left": 0.8995000123977661,
                    "Top": 0.06627999991178513
                }
            }
        }
    ]
}
```

Step 3 – Stopping your model

You incur charges while the Custom Labels model is running. You should stop your model if it's not being used, as shown here:

Figure 3.16: Stopping the model using the Rekognition Custom Labels console

Since this could be a disruptive action in case your model is in use, it will ask you to confirm that you want to stop the model. You should select **Stop**.

Summary

In this chapter, we covered what Amazon Rekognition Custom Labels is and how you can use it to detect objects and scenes in images that are specific to your needs. We discussed the benefits of Rekognition Custom Labels and the process to create a model using the service. In the end, we trained and deployed a model to detect Packt's logo in images.

In the next chapter, we will learn how you can build a contactless hotel check-in system using Amazon Rekognition. You will use different Rekognition features and APIs you have learned about so far in building the check-in system.

Part 2: Applying CV to Real-World Use Cases

This second part consists of three cumulative chapters that will dive deeper into real-world use cases using Amazon Rekognition and other AWS AI services to build applications that demonstrate how to solve business challenges using core CV capabilities.

By the end of this part, you will understand how to incorporate the features of AWS AI services within your identity verification, real-time video analysis, and content moderation workflows.

This part comprises the following chapters:

- *Chapter 4, Using Identity Verification to Build a Contactless Hotel Check-In System*
- *Chapter 5, Automating a Video Analysis Pipeline*
- *Chapter 6, Moderating Content with AWS AI Services*

Using Identity Verification to Build a Contactless Hotel Check-In System

Using facial recognition is one of the most natural ways for humans to identify one another. Our primate eyes instinctively detect metadata, such as the person's age, emotional state, gender, and identity. Numerous business problems are solvable using this information to build personalized experiences.

Las Vegas casinos and hotels are rife with examples. First, they want to greet high rollers and other VIP customers immediately. Second, the Nevada Gaming Commission doesn't allow persons under the age of 21 on the gaming floor. Third, they monitor the emotional state of customers to preemptively route waitstaff and security staff.

Amazon Rekognition's Facial Detection APIs return bounding boxes, landmarks, and pose information for each face it detects. It also predicts age ranges, emotions, gender, and other properties.

In this chapter, you'll learn how to do the following:

- Register customers' profiles using their face
- Authenticate those customers
- Securely update a customer profile
- Register customers using an ID card

Technical requirements

You will require the following:

- Access to an active **Amazon Web Services** (**AWS**) account with permissions to access Amazon SageMaker and Amazon Rekognition

- PyCharm or any Python IDE
- All the code examples for this chapter can be found on GitHub at `https://github.com/PacktPublishing/Computer-Vision-on-AWS`

A Jupyter notebook is available for running the example code from this chapter. You can access the most recent code from this book's GitHub repository, `https://github.com/PacktPublishing/Computer-Vision-on-AWS`. Clone that repository to your local machine using the following command:

```
$ git clone https://github.com/PacktPublishing/Computer-Vision-
on-AWS
$ cd Computer-Vision-on-AWS/04_HotelCheckin
```

Additionally, you will need an AWS account and Jupyter notebook. *Chapter 1* contains detailed instructions for configuring the developer environment.

Prerequisites

To build a contactless casino and resort, you must create a few AWS resources. This section outlines the necessary steps.

Creating the face collection

The contactless hotel and casino needs an Amazon Rekognition collection to hold our face metadata:

```
$ aws rekognition create-collection \
  --region us-east-2 \
  --collection-id "HotelCollection"
```

This command will report the following output:

```
{
  «StatusCode": 200,
  «CollectionArn": "aws:rekognition:region:account:collection/
HotelCollection",
  «FaceModelVersion": "6.0"
}
```

Creating the image bucket

Amazon Rekognition supports analyzing PNG and JPEG images within an S3 bucket. The bucket must reside in the same region as the one used for the Amazon Rekognition public endpoint:

```
$ aws s3api create-bucket \
  --bucket ch04-hotel-use2 \
  --region us-east-2 \
  --create-bucket-configuration \
    LocationConstraint=us-east-2
```

This command will output the following response:

```
{
  «Location»: «http://ch04-hotel-use2.s3.amazonaws.com/»
}
```

> **The InvalidLocationConstraint error**
>
> You may receive the following error creating the bucket in North Virginia (us-east-1). Omit the --create-bucket-configuration parameter to mitigate the issue:
>
> **An error occurred (InvalidLocationConstraint) when calling the CreateBucket operation: The specified location-constraint is not valid.**

Uploading the sample images

This chapter's GitHub repository includes multiple test images for experimenting with the APIs. You must upload them to your Amazon S3 bucket:

```
$ aws s3 sync 04_HotelCheckin/images s3://ch04-hotel-use2/
images
```

The command will print one statement per image uploaded, such as the following:

```
upload: images\Nate-Bachmeier.png to s3://ch04-hotel-use2/
images/Nate-Bachmeier.png
upload: images\Lauren-Mullennex.jpg to s3://ch04-hotel-use2/
images/Lauren-Mullennex.jpg
```

Creating the profile table

You will store the customer profile information inside Amazon DynamoDB, a serverless NoSQL database. It offers a simple interface to persist and retrieve values based on object keys:

```
$ aws dynamodb create-table \
  --region us-east-2 \
  --table-name HotelProfile \
```

```
--attribute-definitions \
  AttributeName=PartitionKey,AttributeType=S \
  AttributeName=SortKey,AttributeType=S \
--key-schema \
  AttributeName=PartitionKey,KeyType=HASH \
  AttributeName=SortKey,KeyType=RANGE \
--billing-mode PAY_PER_REQUEST
```

Executing this command will output the following table structure and configuration information:

```
{
    «TableDescription": {
        «AttributeDefinitions": [
            {
                «AttributeName": "PartitionKey",
                «AttributeType": "S"
            },
            {
                «AttributeName": "SortKey",
                «AttributeType": "S"
            }
        ],
        «TableName": "HotelProfile",
        «KeySchema": [
            {
                «AttributeName": "PartitionKey",
                «KeyType": "HASH"
            },
            {
                «AttributeName": "SortKey",
                «KeyType": "RANGE"
            }
        ],
        «TableStatus": "CREATING",
        «CreationDateTime": "2022-12-11T14:31:35.401000-05:00",
        «ProvisionedThroughput": {
            «NumberOfDecreasesToday": 0,
            «ReadCapacityUnits": 0,
```

```
      «WriteCapacityUnits": 0
    },
    «TableSizeBytes": 0,
    «ItemCount": 0,
    «TableArn": "arn:aws:dynamodb:region:accountid:table/
HotelProfile",
    «TableId": "f23ec62c-081f-4234-93de-af6fae2e8a8d",
    «BillingModeSummary": {
      «BillingMode": "PAY_PER_REQUEST"
    }
  }
}
```

Introducing collections

Amazon Rekognition stores facial information inside server-side containers known as *collections*. Collections represent a logical grouping of Face Metadata, not the original image of the person. It supports operations for indexing, listing, searching, and deleting faces.

There's no charge for collections; each holds 10 million Face Metadata, and you only pay for the aggregate Faces Metadata Storage ($0.00001/per facial metadata per month). For example, storing one million faces costs $1 per month, regardless of the total spanning collections.

Creating a collection

You can create a collection using the **AWS Command Line Interface** (**AWS CLI**) v2 or the AWS SDK. When you invoke the CreateCollection API, it only requires a name and returns instantly:

```
$ aws rekognition create-collection --collection-id \
"HelloWorld"
```

This command will report the following output:

```
{
  «StatusCode": 200,
  «CollectionArn": "aws:rekognition:us-east-
2:1234567890:collection/HelloWorld",
  «FaceModelVersion": "6.0"
}
```

The following table enumerates the `CreateCollection` response's properties:

Property	Type	Description
CollectionArn	String	The **Amazon Resource Name** (**ARN**) of the collection
FaceModelVersion	String	The version number of the facial detection model
StatusCode	Integer	The HTTP status code of the operation

Table 4.1: The CreateCollection response

Describing a collection

Next, you can inspect the collection using the `DescribeCollection` API:

```
aws rekognition describe-collection --collection-id \
"HelloWorld"
```

This command will output the following response:

```
{
    «FaceCount": 0,
    «FaceModelVersion": "6.0",
    «CollectionARN": "arn:aws:rekognition:us-east-2:
1234567890:collection/HelloWorld",
    «CreationTimestamp": 1670781712.443
}
```

The following table enumerates the `DescribeCollection` response's properties:

Property Name	Type	Description
CollectionArn	String	The ARN of the collection
FaceModelVersion	String	The version number of the facial detection model
FaceCount	Integer	The number of faces indexed in the collection
CreationTimestamp	Timestamp	The number of milliseconds from the Unix epoch time (January 1, 1970)

Table 4.2: The DescribeCollection response

Deleting a collection

Finally, you can use the `DeleteCollection` API to remove the collection:

```
aws rekognition delete-collection --collection-id \
"HelloWorld"
```

This command will output the following response:

```
{
   «StatusCode": 200
}
```

The `DeletectCollection` response only contains the HTTP status code. You don't need to check the status when calling this API from Python, as boto3 throws an exception for error conditions:

Property	Type	Description
StatusCode	Integer	The HTTP status code of the operation

Table 4.3: The DeleteCollection response

Quick recap

In this section, you learned that collections are containers that can hold millions of facial metadata objects. Amazon Rekognition provides APIs for creating, listing, describing, and deleting containers. Next, you'll use these resources to store and query persons within images.

Describing the user journeys

There are four distinct user journeys for our contactless casino and resort: registering by face, registering by ID card, authenticating the user, and updating the user's profile.

The first journey requires collecting metadata and registering the customer's profile. You will need to automate steps for checking image quality, confirming uniqueness, and persisting profile state.

Registering a new user

The first step to registering a new user is to validate that the image meets your requirements. Amazon Rekognition's `DetectFaces` APIs provide the building blocks for automating these checks, such as the person looking into the camera.

Next, you'll need to search the Amazon Rekognition collection and determine whether this is a repeat request or new registration. Using the `SearchFacesByImages` API handles this complexity for you.

After confirming that the incoming face is unique, the `IndexFaces` API will securely persist the Face Metadata into the collection.

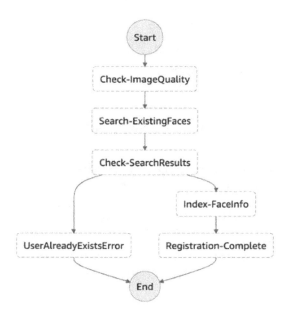

Figure 4.1: Registering a new user process

Authenticating a user

Once the customer registers, you can quickly authenticate them using an image. This user journey begins with verifying the incoming image's quality.

Next, use the `CompareFaces` API to assess the similarity of that incoming image against a cached copy. This operation is less expensive and more performant than searching the collection every time.

Suppose you have a cache miss. In that case, use the `SearchFacesByImage` API to find the expected face and grant the authentication using the similarity scores.

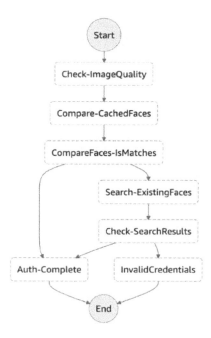

Figure 4.2: Authenticating a user process

Registering a new user with an ID card

Next, a more sophisticated version of registering a new user journey is with a government ID card. Many organizations want customers and employees to register through a self-service portal.

This chapter covers integrating Amazon Textract to read a person's government ID card. You will also leverage the CompareFaces API to confirm the ID card's picture matches the registered user.

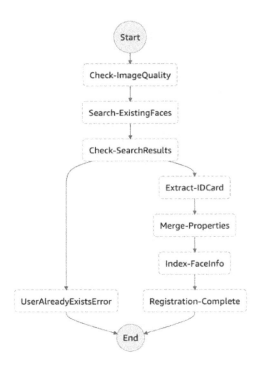

Figure 4.3: Registering a user with an ID card process

Updating the user profile

The fourth user journey updates the user's profile and security information. Like the authentication flow, you should use the CompareFaces API and fall back to the SearchFacesByImage API in error conditions.

After verifying the user's identity, you call the IndexFaces API to override the previous state information.

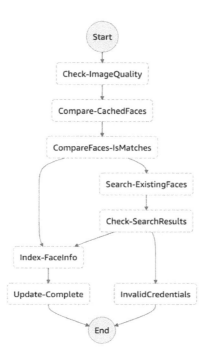

Figure 4.4: Update user profile process

At this point, you should have a high-level understanding of the steps needed to perform the four user journeys. Now let's deep dive into the code fragments necessary to implement them.

Implementing the solution

You probably noticed that multiple user flows, such as `Check Image Quality`, contain the same tasks. We can encapsulate that logic into Python functions (or AWS Lambda functions) to avoid code duplication.

Checking image quality

The first step in every user journey is to confirm that the photograph is usable. For our use case, this means the photo contains one person, and they're looking at the camera.

The `DetectFaces` API assesses an image and returns the 100 more prominent faces. You'll receive high confidence scores for frontal faces and notice performance degradations with photos from obscure angles.

Like other Amazon Rekognition APIs, you must specify either a base64-encoded image or a location within an Amazon S3 bucket:

```python
import boto3

region_name = 'us-east-2'
bucket_name = 'ch04-hotel-use2'
image_name = 'images/Nate-Bachmeier.png'

rekognition = boto3.client('rekognition',region_name=region_
name)

faces = rekognition.detect_faces(
    Attributes=['ALL'],
    Image={
     'S3Object':{
        'Bucket': bucket_name,
        'Name': image_name
      }
    })
```

After running that snippet, you can inspect the `faces` variable to review the various predictions:

```python
def print_faces(faces):
    counter=0
    for face in faces[<FaceDetails']:
        counter+=1
        age_range = face['AgeRange']
        gender = face[<Gender>]
        emotions = face[<Emotions>]

        age = (age_range['High'] + age_range['Low'])/2
        emotions.sort(key=lambda x: x['Confidence'],
reverse=True)
        top_emotion = emotions[0]

        print('Person %d is %s(%2.2f%%) around %d age and
%s(%2.2f%%) state.' % (
            counter,
```

```
                gender[<Value>],
                gender[<Confidence>],
                age,
                top_emotion['Type'],
                top_emotion['Confidence']
        ))


print_faces(faces)
```

You should see an output similar to the following example:

Person 1 is Male(100.00%) around 31 age and CALM(96.24%) state.

The following table enumerates the DetectFaces response's properties. For most scenarios, you'll only use the FaceDetails property. Amazon Rekognition doesn't perform image correction for JPEG and PNG files that don't contain Exif metadata:

Property Name	Type	Description
FaceDetails	Array of FaceDetail	Provides the details of each detected face within the image
OrientationCorrection	String	The value of OrientationCorrection is always null

Table 4.4: The DetectFaces response

Using the FaceDetails property, you can codify the requirements for the check-in registration system. Specifically, we want one person in the image, facing forward, without sunglasses, and to be well lit:

```
def has_only_one_face(faces):
    if len(faces['FaceDetails']) == 1:
        return True
    return False


def is_facing_forward(face):
    for dimension in [<Pitch','Roll','Yaw']:
        value = face[<Pose'][dimension]
        if not (-45 < value and value < 45):
```

```
            return False
        return True

    def has_sunglasses(face):
        sunglasses = face[<Sunglasses'] ['Value']
        return sunglasses

    def is_well_lit(face):
        if face[<Quality'] ['Brightness'] < 25:
            return False
        return True
```

Finally, we can implement a policy check using these utility functions:

```
    def check_faces(faces):
        if not has_only_one_face(faces):
            print('Incorrect face count.')
            return False

        user_face = faces['FaceDetails'] [0]
        if not is_facing_forward(user_face):
            print('Customer not facing forward')
            return False

        if has_sunglasses(user_face):
            print('Please take off sunglasses.')
            return False

        if not is_well_lit(user_face):
            print('The image is blurry')
            return False

        print("Valid face detected!")
        return True
```

The following table enumerates the FaceDetails object's properties. Several properties contain Boolean Value and Confidence scores of the detection. For example, Beard is either present or not. When the Confidence score is too low, you should consider it false or erroneous:

Property Name	Description
AgeRange	The estimated face's age range in years
Beard	Indicates whether the face has a beard
BoundingBox	The bounding box around the face within the image
Emotions	An array of emotional state predictions
Eyeglasses	Indicates whether the face has eyeglasses
EyesOpen	Indicates whether the face has eyes open
Gender	Predicts the face is male or female
Landmarks	Indicates the X and Y coordinates of core facial features (such as eyes, ears, and nose)
MouthOpen	Indicates whether the face has an open mouth
Mustache	Indicates whether the face has a mustache
Pose	Indicates the face's pitch, roll, and yaw
Quality	Indicates image brightness and sharpness
Smile	Indicates whether the face is smiling
Sunglasses	Indicates whether the face is wearing sunglasses

Table 4.5: The FaceDetail object

Indexing face information

After validating that the image meets your requirements, you'll want to use the `IndexFaces` API to persist facial metadata into an Amazon Rekognition collection.

During this operation, specify `ExternalImageId` to tag all detected faces. For instance, this snippet tags the face with the author's alias:

```
collection_id="HotelCollection"
image_set = {
    'nbachmei': 'images/Nate-Bachmeier.png',
    <lemull: 'images/Lauren-Mullennex.jpg'
}


for (externalImageId, object_name) in image_set.items():
    rekognition.index_faces(
        CollectionId = collection_id,
```

```
        ExternalImageId=externalImageId,
        Image={
            <S3Object':{
                <Bucket>: bucket_name,
                <Name>: object_name
            }
        })
```

The following table enumerates the `IndexFace` parameters:

Parameter Name	Required	Type	Description
CollectionId	Yes	String	The targeted Amazon Rekognition collection's name
DetectionAttributes	No	String Array	Specifies which properties to populate into the collection
ExternalImageId	No	String	The ID you want to assign to all the faces detected in the image
Image	Yes	Image	The image to index
MaxFaces	No	Integer	The maximum number of faces to index (1 to 100)
QualityFilter	No	String	A control to prevent indexing blurry faces

Table 4.6: The IndexFaces parameters

Search existing faces

You retrieve facial metadata from a collection using the `SearchFacesByImage` API. This action detects and uses only the most prominent face in the input image:

```
search_image = 'images/SearchFacesByImageExample.jpg'

response = rekognition.search_faces_by_image(
    CollectionId=collection_id,
    Image={
        <S3Object':{
            <Bucket>: bucket_name,
            <Name>: search_image
```

```
        }
    })
```

Next, let's inspect the search results returned:

```
def print_search_results(search_response):
    for match in search_response['FaceMatches']:
        externalImageId = match['Face']['ExternalImageId']
        confidence = match[<Face'][' Confidence']

        print('This image is %s (%2.2f%% confidence)' % (
            externalImageId,
            confidence
        ))


print_search_results(response)
```

The previous snippet will output a line similar to this output. You'll notice that it contains `ExternalImageId` specified during the `Index Face` API:

This image is nbachmei (100.00% confidence)

Quick recap

In this section, you confirmed that an image is usable for your scenario, such as one person in the picture without sunglasses. These simple-to-use properties open the door to building more complex policies. Next, you added the facial metadata into a collection and retrieved that identity using another image.

Finally, let's put this user flow's pieces together. The `FacialDetection.py` file in this chapter's repository does precisely that. Next, let's review and add capabilities for registering users with government identification cards.

Supporting ID cards

A standard requirement for contactless registration systems is to import the person's government identification card, such as a driver's license. We can add this capability using Amazon Textract and Amazon Rekognition.

Reading an ID card

Amazon Textract is a serverless service that automatically extracts text, handwriting, and data from scanned documents. In addition, it can detect forms and tables within documents, PDFs, and images.

In this section, you'll use Amazon Textract to read a government card and report the properties. There are three steps to completing this task:

1. Create the Amazon Textract client.

2. Use the Amazon Textract API.

3. Convert the response into a simple dictionary.

First, you must initialize the Amazon Textract client. Luckily, creating an Amazon Textract client is nearly identical to the process for Amazon Rekognition:

```
textract = boto3.client('textract',region_name=region_name)
```

The Amazon Textract API contains actions for analyzing various document types. We will use `AnalyzeID` as its purpose-built for this scenario. Suppose your application needs to extract text from additional documents – in that case, you can repeat this process to call those APIs:

API Name	Description
AnalyzeDocument	Discovers tables, forms, paragraphs, and signatures within a document
AnalyzeExpense	Discovers line items and invoice summary information within a document
AnalyzeID	Discovers key/value pairs within identity documents

Table 4.7: The Amazon Textract Analyze API

To call the `AnalyzeID` API, you'll need to specify the `DocumentPages` array as one or two elements representing the identification card's front and back. The local `passport_card` variable references a file from this chapter's repository. You uploaded this file during the *Prerequisite* section's *Upload the sample images* task:

```
passport_card = 'images/passport_card.jpeg'

response = textract.analyze_id(
    DocumentPages=[
        {
            «S3Object":{
                «Bucket»: bucket_name,
                «Name»: passport_card
            }
        }
    ])
```

The following table enumerates the AnalyzeID response's properties. You can determine the total Pages processed from DocumentMetadata. Any findings are available from the IdentityDocuments array, which contains one element per DocumentPages item:

Property Name	Type	Description
AnalyzeIDModelVersion	String	The version of the AnalyzeIdentity API used to process the document
DocumentMetadata	DocumentMetadata	Information about the input document
IdentityDocuments	Array of IdentityDocument	The list of documents processed by the AnalyzeID API

Table 4.8: The AnalyzeID response

Next, inspect the response variable to get the ID card's properties:

```
fields = {}
for document in response['IdentityDocuments']:
    for field in document[<IdentityDocumentFields']:
        key = field[<Type'][' Text']
        value = field[<ValueDetection'][' Text']
        fields[key] = value

from json import dumps
```

```
print(dumps(fields,indent=2))
```

You should see an output similar to this JSON representation:

```
{
    «FIRST_NAME»: «HAPPY»,
    «LAST_NAME»: «TRAVELER»,
    «MIDDLE_NAME»: «»,
    «SUFFIX»: «»,
    «CITY_IN_ADDRESS»: «»,
    «ZIP_CODE_IN_ADDRESS»: «»,
    «STATE_IN_ADDRESS»: «»,
    «STATE_NAME»: «»,
    «DOCUMENT_NUMBER»: «C03005988»,
    «EXPIRATION_DATE»: «29 NOV 2019»,
    «DATE_OF_BIRTH»: «1 JAN 1981»,
    «DATE_OF_ISSUE»: «30 NOV 2009»,
    «ID_TYPE»: «DRIVER LICENSE FRONT»,
    «ENDORSEMENTS»: «»,
    «VETERAN»: «»,
    «RESTRICTIONS»: «»,
    «CLASS»: «»,
    «ADDRESS»: «»,
    «COUNTY»: «»,
    «PLACE_OF_BIRTH»: «NEW YORK U.S.A.»,
    «MRZ_CODE»: «»
}
```

The IdentityDocument also has properties for inspecting the document's structure. For example, you can query the Blocks property to find the relationship between different words, lines, and tables, among other types. Like Amazon Rekognition and other AWS AI services, the response's child objects contain the Confidence scores that should influence whether you accept a given field's predicted value:

Property Name	Type	Description
`Blocks`	Array of Block	Individual word recognition.
`DocumentIndex`	Integer	Denotes the placement of a document in the `IdentityDocument` list. The first document is marked one.
`IdentityDocumentFields`	Array of `IdentityDocumentField`	List of the discovered identity fields and values.

Table 4.9: The IdentityDocument object

Using the CompareFaces API

When applications can read customers' ID cards, it reduces tedious data entry. But how do you know the card belongs to that person? In the physical world, we look at the picture and compare it to the person. Amazon Rekognition offers a similar capability through its `CompareFaces` API. To use this feature, you specify `SourceImage` and `TargetImage`.

Let's test this functionality by checking whether two passport cards contain this chapter's author. You will need to use the following object in your Amazon S3 bucket:

```
nbachmei_image = "images/Nate-Bachmeier.png"
match_passport = "images/nbachmei-passport.jpg"
nonmatch_passport = "images/passport_card.jpeg"
```

Next, define the compare_faces_test function to run the comparison and print the similarity score:

```
def compare_faces_test(passport_card):
    comparison_match = rekognition.compare_faces(
        SourceImage={
            "S3Object":{
                "Bucket": bucket_name,
                "Name": nbachmei_image
            }
        },
        TargetImage={
            "S3Object":{
```

```
                «Bucket»: bucket_name,
                «Name»: passport_card
            }
        })

    if len(comparison_match['FaceMatches'])  > 0:
        for face in comparison_match['FaceMatches']:
            similarity = face[<Similarity>]
            print('%s face with %2.2f%% similarity ' % (
                passport_card,
                similarity))
    else:
        print("%s has %d unmatched faces" % (
            passport_card,
            len(comparison_match['UnmatchedFaces']))
        )
```

Finally, you can run the test cases using these two lines:

```
compare_faces_test(match_passport)
compare_faces_test(nonmatch_passport)
```

The commands will print output similar to the following:

```
images/nbachmei-passport.jpg face with 98.77% similarity
images/passport_card.jpeg has 1 unmatched faces
```

Quick recap

In this section, you used Amazon Textract to read an ID card and Amazon Rekognition to verify that it belongs to the holder. These capabilities help streamline your data entry for user registration workflows. You could incorporate them into authentication flows as a **Multi-Factor Authentication (MFA)** scheme, such as checking an employee badge.

Now that you've explored the Face APIs and understand how to implement them, it's time to see them in action. Let's deploy **Rekognition Identity Verification (RIV)**, an open source web application, using these patterns.

Guidance for identity verification on AWS

The AWS Solutions Architecture team has built a complete code sample of this topic, including a ReactJS-based web interface. You can download the most recent version from `https://aws.amazon.com/solutions/guidance/identity-verification-on-aws/`.

Solution overview

The solution can be represented as follows:

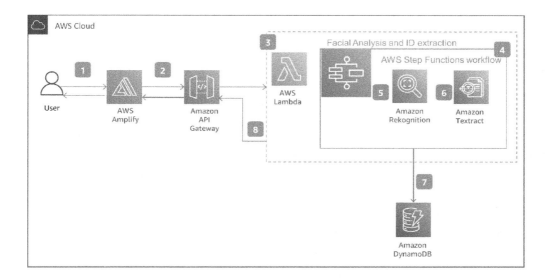

Figure 4.5: Overview of the workflow

This workflow can be broken down into the following steps:

1. Customers can use their web camera to upload images and ID cards through the web interface.
2. Amplify routes the request through Amazon API Gateway.
3. API Gateway invokes AWS Lambda functions to analyze the submitted image.
4. AWS Step Functions orchestrate the four user journeys.
5. Amazon Rekognition analyzes image data.
6. Amazon Textract analyzes any ID card requests.
7. The results persist in an Amazon DynamoDB table.

Deployment process

Choose the appropriate CloudFormation stack to create the solution in your AWS account in your preferred AWS Region. This solution deploys API Gateway integrated with Step Functions and Amazon Rekognition APIs to run the identity verification workflows.

Choosing on one of the following launch buttons will launch the solution into your AWS account in the particular region.

For additional deployment options and troubleshooting tips, see *Accelerate your identity verification projects using AWS Amplify and Amazon Rekognition sample implementations* (`https://aws.amazon.com/blogs/machine-learning/accelerate-your-identity-verification-projects-using-aws-amplify-and-amazon-rekognition-sample-implementations/`).

Cleanup

To prevent accruing additional charges on your AWS account, delete the resources you provisioned by navigating to the AWS CloudFormation console and deleting the `Riv-Prod` stack.

Deleting the stack doesn't delete the S3 bucket you created, and this bucket stores all the face images. If you want to delete the S3 bucket, navigate to the Amazon S3 console, empty it, and then confirm that you want to delete it permanently.

Summary

Countless business processes rely on identifying and classifying humans. In this chapter, you learned how Amazon Rekognition makes it easy to automate those processes using facial metadata.

For example, you can quickly add capabilities to predict age, gender, and emotional state to your applications. These capabilities enable scenarios such as detecting underage customers on a casino gaming floor. You can also preemptively notice which customers deserve more attention or are becoming hostile.

You learned about bringing these various capabilities together by building a contactless identity verification system. That solution includes building blocks for registering, authenticating, and updating customers' profiles. Then you added support for corporate and government identification cards using Amazon Textract.

In the next chapter, you'll learn how to automate a video analysis pipeline using IP cameras and OpenCV. That solution demonstrates some of the many scenarios that Amazon Rekognition and AWS Amplify.

5

Automating a Video Analysis Pipeline

If a picture is worth a thousand words, what's the value of a video? In the previous chapters, you've learned how to use Amazon Rekognition for image analysis. Now it's time to introduce some techniques for handling video content.

You'll collect frames from IP cameras and analyze them in the cloud. This task leverages the **Real Time Streaming Protocol (RTSP)**, OpenCV, and Amazon Rekognition. Next, using the Amazon Rekognition Video's Person Tracking API, you'll extract walkways from video feeds. Furthermore, the demonstrated design patterns are serverless and elastically scale to virtually any size!

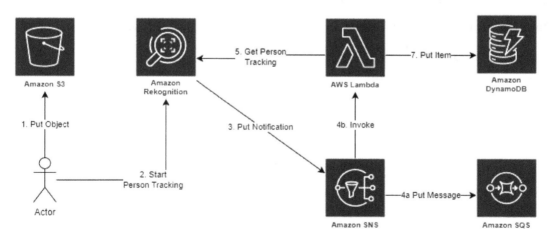

Figure 5.1: Target state architecture

In this chapter, you'll learn how to do the following:

- Sample IP camera frames using OpenCV
- Build an event-based analysis pipeline
- Publish custom Amazon CloudWatch metrics
- Track people's paths within stored video clips

Technical requirements

A Jupyter notebook is available for running the example code from this chapter. You can access the most recent code from this book's GitHub repository, https://github.com/PacktPublishing/Computer-Vision-on-AWS. Clone that repository to your local machine using the following command:

```
$ git clone https://github.com/PacktPublishing/Computer-Vision-on-AWS
$ cd Computer-Vision-on-AWS/05_VideoAnalysis
```

Creating the video bucket

Amazon Rekognition supports analyzing MOV and MP4 videos within an S3 bucket. The bucket must reside in the same AWS Region as the Amazon Rekognition endpoint used.

Bucket names must be globally unique, so you must modify the --bucket parameter:

```
$ aws s3api create-bucket \
  --bucket ch05-video-use2 \
  --region us-east-2 \
  --create-bucket-configuration LocationConstraint=us-east-2
```

This command will output the following response:

```
{
    "Location": "http://ch05-video-use2.s3.amazonaws.com/"
}
```

> InvalidLocationConstraint error
>
> You might receive the following error creating the bucket in North Virginia (us-east-1). Omit the `--create-bucket-configuration` parameter to mitigate the issue:
>
> ```
> An error occurred (InvalidLocationConstraint) when calling the
> CreateBucket operation: The specified location-constraint is
> not valid.
> ```

Uploading content to Amazon S3

This chapter's repository contains sample video recordings for completing the material. You must upload these files into an Amazon S3 bucket in the same AWS Region as the Amazon Rekognition endpoint:

```
$ aws s3 sync 05_VideoAnalysis s3://ch05-video-use2
```

Creating the person-tracking topic

The `PersonTracking` API requires an Amazon **Simple Notification Service (SNS)** topic to receive asynchronous updates:

```
aws sns create-topic --name
AmazonRekognitionPersonTrackingTopic --region us-east-2
```

You should see an output similar to the following. The response's `region` and `accountid` values will be unique to your AWS account:

```
{
    "TopicArn": "arn:aws:sns:region:accountid:
AmazonRekognitionPersonTrackingTopic"
}
```

Assign the `TopicArn` property to a local variable in your terminal:

```
$ TOPIC_ARN= arn:aws:sns:region:accountid:
AmazonRekognitionPersonTrackingTopic
```

Subscribing a message queue to the person-tracking topic

You'll need a mechanism for capturing the topic messages for offline analysis. An easy solution is subscribing to an Amazon **Simple Queuing Service (SQS)** message queue.

First, create the message queue in the same AWS Region as the topic:

```
aws sqs create-queue \
  --queue-name PersonTrackingQueue \
  --region us-east-2
```

You should see an output similar to this:

```
{
    "QueueUrl": "https://region.queue.amazonaws.com/accountid/
PersonTrackingQueue"
}
```

Assign the QueueUrl property to a local variable in your terminal:

```
$ QUEUE_URL= https://region.queue.amazonaws.com/accountid/
PersonTrackingQueue
```

Next, look up the queue's **Amazon Resource Name (ARN)**:

```
aws sqs get-queue-attributes \
  --region us-east-2 \
  --attribute-names QueueArn \
  --queue-url $QUEUE_URL
```

You should see an output similar to this example:

```
{
    "Attributes": {
        "QueueArn":
"arn:aws:sqs:region:account:PersonTrackingQueue"
    }
}
```

Assign the QueueArn property to a local variable in your terminal:

```
$ QUEUE_ARN=arn:aws:sqs:region:account:PersonTrackingQueue
```

Third, pass TopicArn and QueueArn to the Amazon SNS Subscribe API:

```
aws sns subscribe \
  --region us-east-2 \
```

```
--topic-arn $TOPIC_ARN \
--protocol sqs \
--notification-endpoint $QUEUE_ARN
```

You should see an output similar to this example:

```
{
    "SubscriptionArn": "arn:aws:sns:region:account:AmazonRekogn
itionPersonTrackingTopic:04877b15-7c19-4ce5-b958-969c5b9a1ecb"
}
```

Finally, grant the topic permission to write to the message queue. Create a local file called PersonTrackingQueuePolicy.json with the following content:

> **Security best practices**
>
> The following example uses wildcards to be more flexible within your environment. For production systems, its best practice to avoid wildcards and restrict access to specific resources.

```
{
    "Statement": [
      {
        "Effect": "Allow",
        "Principal": {
          "Service": "sns.amazonaws.com"
        },
        "Action": "sqs:SendMessage",
        "Resource": "arn:aws:sqs:*:*:*PersonTracking*",
        "Condition": {
          "ArnEquals": {
            "aws:SourceArn": "arn:aws:sns:*:*:*PersonTracking*"
          }
        }
      }
    ]
}
```

You'll need to slash-encode the file for the AWS CLI to pass it to the `SetQueueAttributes` API correctly:

```
{
    "Policy": "{\n      \"Statement\":
[\n         {\n           \"Effect\":
\"Allow\",\n              \"Principal\": {\n                \"Service\":
\"sns.amazonaws.com\"\n             },\n          \"Action\":
\"sqs:SendMessage\",\n          \"Resource\":
\"arn:aws:sqs:*:*:*PersonTracking*\",\n             \"Condition\":
{\n           \"ArnEquals\": {\n               \"aws:SourceArn\":
\"arn:aws:sns:*:*:*PersonTracking*\"\n            }\n          }\n
    }\n     ]\n}"
}
```

After correctly encoding the file, you apply it using the following command:

```
$ aws sqs set-queue-attributes \
   --region us-east-2 \
   --queue-url $QUEUE_URL \
   --attributes file://PersonTrackingQueuePolicy.json
```

Creating the person-tracking publishing role

Amazon Rekognition Video's `StartPersonTracking` API requires an Amazon SNS topic and AWS **Identity and Access Management** (**IAM**) role with sufficient permissions to post that notification.

First, you must create the `PersonTrackingAssumeRole.json` document for Amazon Rekognition:

```
{
    "Version": "2012-10-17",
    "Statement": [
        {
            "Effect": "Allow",
            "Principal": {
                "Service": "rekognition.amazonaws.com"
            },
            "Action": "sts:AssumeRole",
            "Condition": {}
        }
```

```
        ]
    }
```

Next, associate this policy document with a new service role:

```
$ aws iam create-role \
  --role-name PersonTrackingPublisher \
  --assume-role-policy-document
    file://PersonTrackingAssumeRole.json
```

This command will output the role's description similar to the following:

```
{
    "Role": {
        "Path": "/",
        "RoleName": "PersonTrackingPublisher",
        "RoleId": "AROAYOW64NPHJUINSJXWQ",
        "Arn": "arn:aws:iam::accountid:role/
PersonTrackingPublisher",
        "CreateDate": "2022-12-26T04:01:49Z",
        "AssumeRolePolicyDocument": {
            "Version": "2012-10-17",
            "Statement": [
                {
                    "Effect": "Allow",
                    "Principal": {
                        "Service": "rekognition.amazonaws.com"
                    },
                    "Action": "sts:AssumeRole",
                    "Condition": {}
                }
            ]
        }
    }
}
```

Finally, associate the managed policy, `AmazonRekognitionServiceRole`:

```
aws iam attach-role-policy \
  --role-name PersonTrackingPublisher \
  --policy-arn arn:aws:iam::aws:policy/service-role/
AmazonRekognitionServiceRole
```

> **Custom topic name permissions**
>
> The `AmazonRekognitionServiceRole` managed policy can only write to Amazon SNS topics whose name begins with `AmazonRekognition`. If your custom topic has a different prefix, the `PersonTrackingPublisher` role requires an inline policy to grant access.
>
> Non-production systems can also mitigate this issue by attaching the policy ARN `arn:aws:iam::aws:policy/AmazonSNSFullAccess`. This policy grants the role full access to all Amazon SNS topics within the account.

Setting up IP cameras

This chapter's core objective is tracking a pet's movements through the home using IP cameras. I choose eufyCam 2C Pro cameras because they capture 2K video, don't require a subscription, and offer RTSP endpoints. RTSP is an industry standard for security cameras and many video-sharing technologies.

If you don't have a security camera, don't panic. This chapter's repository contains artifacts for completing the *Using the PersonTracking API* section.

> **Choosing different camera models**
>
> If you select a different brand or model, confirm that it supports RTSP. Use your preferred search engine to look for `camera model and rtsp` to find the most relevant information.

You broker access to the cameras using the Eufy base station's RTSP endpoints. There's one endpoint per camera starting address: `rtsp://base-station-ip/live0`.

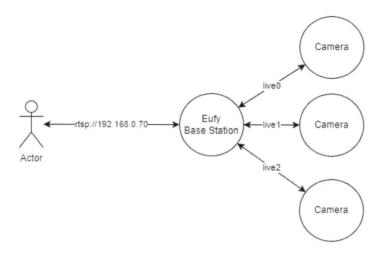

Figure 5.2: Camera base station configuration

Download Gardinal's free RTSP Player (`https://rtspplayer.com`) to confirm the endpoint is accessible. If RTSP Player cannot access the endpoint, confirm the following:

1. The base station has RTSP protocol enabled.
2. The base station IP address is correct.
3. The camera is active (some models are motion activated).

Figure 5.3: RTSP Player

Quick recap

In this section, you configured the prerequisite AWS resources, including a bucket, topic, and queue. You'll need these resources to complete the *Using the GetPersonTracking API* section later. Additionally, you should be able to connect to your IP camera using the RTSP protocol. With everything set up and ready, let's use CV for video analysis.

Using IP cameras

After confirming that the RTSP endpoint(s) are accessible, it is time to connect using the open source computer vision library **OpenCV**. OpenCV for Python is a collection of tools and capabilities for interacting with image and video content. It contains a ton of functionality, enough to fill multiple books. However, we'll only scratch the surface and use it to collect frames from our cameras.

Installing OpenCV

OpenCV's Python installation has several platform-dependent binaries, so you must be mindful of where the Python code will execute. This chapter uses an x86 64bit Amazon Linux 2 machine:

```
$ yum -y update && yum -y install \
  mesa-libGL.x86_64 \
  opencv-python.x86_64 \
  python3
```

Suppose you're using an Apple M1 or Amazon Graviton processor. In that case, you must specify the ARM64 platform:

```
$ yum -y update && yum -y install \
  mesa-libGL.arm64 \
  opencv-python.arm64 \
  python3
```

Installing additional modules

Next, you must install the following Python modules:

```
pip3 install rtsp
pip3 install boto3
```

Connecting with OpenCV

Open this chapter's Juypter notebook and set the base station's configuration to match your home setup. For instance, my base station is at 192.168.0.70 and requires a username of admin with the password EYE_SEE_YOU.

The cameras variable maps the endpoint name to a room containing the camera. The dictionary values can be any value up to 256 Unicode characters in length:

```
from rtsp import Client
base_station = 'rtsp://admin:EYE_SEE_YOU@192.168.0.70'
cameras = {
    'live0':'office',
    'live1':'kitchen',
    'live2':'living_room',
}
```

Next, you'll want to iterate through the configuration and attempt to fetch an active frame from the cameras. This process starts with connecting to each camera endpoint and confirming whether it is accessible.

When you connect to the base station, it must proxy the session to the camera. This double hop can result in the first few frames being empty. A simple solution is to sleep for 100ms until the stream is ready:

```
from time import sleep
def get_frame(base_station, endpoint):
    rtsp_server_uri = '%s/%s' % (base_station, endpoint)
    with Client(rtsp_server_uri=rtsp_server_uri, verbose=False) as client:
        if not client.isOpened():
            print('{} server is down.'.format(rtsp_server_uri))
            return None
        else:
            print('{} server is up.'.format(rtsp_server_uri))

        while True:
            image = client.read()
            if image is None:
                sleep(0.100)
            else:
                return image
```

You can run this function across all cameras using the following snippet:

```
frames= dict()
for endpoint in cameras.keys():
    frames[endpoint]= get_frame(base_station,endpoint)
```

After the code completes, it reports output similar to the following example. You can read this output as saying that only live0 is active, and the other cameras are offline. The online camera returned an RGB image that's 1920x1080 pixels in size:

```
rtsp://admin:EYE_SEE_YOU@192.168.0.70/live0 server is up.
rtsp://admin:EYE_SEE_YOU@192.168.0.70/live1 server is down.
rtsp://admin:EYE_SEE_YOU@192.168.0.70/live2 server is down.
rtsp://admin:EYE_SEE_YOU@192.168.0.70/live3 server is down.
{'live0': <PIL.Image.Image image mode=RGB size=1920x1080>,
'live1': None, 'live2': None, 'live3': None}
```

Viewing the frame

The get_frame function returns a **Python Imaging Library** (PIL) image object. You might recall using PIL in *Chapter 2* for drawing bounding boxes. In addition, PIL can read and write image data in various formats:

```
frames['live0'].save('images/live0.png', format='PNG')
```

You can also use the PIL image's show method to write the image data into the temp folder and launch the default viewing tool:

```
frames['live0'].show()
```

Figure 5.4: Captured frame from the live0 endpoint

Uploading the frame

Now that you can fetch frames from the IP cameras, the next task is to upload them into Amazon S3. This process requires creating the `boto3` client:

```
import boto3
region_name = 'us-east-2'
s3 = boto3.client('s3', region_name=region_name)
bucket = 'ch05-video-use2'
```

Then you must convert the PIL image into a binary image file:

```
from io import BytesIO
frame_bytes = BytesIO()
frames['live0'].save(frame_bytes, format='PNG')
```

Finally, invoke the `PutObject` API to upload the file to a date-time partitioned folder structure. You can specify up to 10 object tags using the `Metadata` property. This approach helps carry any additional context forward:

```
from datetime import datetime
dt = datetime.now()
object_key = 'frames/%s/%s.png' %(
    cameras['live0'],
```

```
    dt.strftime('%Y/%m/%d/%H/%M/%S.%f')
)
response = s3.put_object(
    Bucket=bucket,
    Key=object_key,
    Body=frame_bytes.getvalue(),
    Metadata={
        'Camera': cameras['live0'],
    })
```

If any issues occur during the file upload, the PutObject API will raise an exception. You can optionally inspect the response for more information:

```
{
    "ResponseMetadata": {
        "RequestId": "ZAGZ8CJGEY28BJ18",
        "HostId": "jdf5IN7y4hl52BJx4a1/
iWKdBQ2hwW9FVuuKgTVZW424q+Ud5EU+LY7kjkE2ku6ZTkZtum1WBlY=",
        "HTTPStatusCode": 200,
        "HTTPHeaders": {
            "x-amz-id-2": "jdf5IN7y4hl52BJx4a1/
iWKdBQ2hwW9FVuuKgTVZW424q+Ud5EU+LY7kjkE2ku6ZTkZtum1WBlY=",
            "x-amz-request-id": "ZAGZ8CJGEY28BJ18",
            "date": "Sun, 25 Dec 2022 22:10:45 GMT",
            "etag": "\"7bddaa72bff30f00441e7fa3df9719f2\"",
            "server": "AmazonS3",
            "content-length": "0"
        },
        "RetryAttempts": 0
    },
    "ETag": "\"7bddaa72bff30f00441e7fa3df9719f2\""
}
```

Reporting frame metrics

As you add more cameras, monitoring which ones emit images becomes challenging. You can implement per-camera performance counters using Amazon CloudWatch custom metrics.

The `increment_frame_count` function adds one to our custom `VideoAnalysis FrameCount` metric by default. Amazon CloudWatch will maintain a different metric instance for each unique combination of dimensions.

Production applications should batch metric data points into a single `PutMetricData` request. Combining up to 1 MB and 1,000 data points reduces the risk of Amazon CloudWatch throttling metric data:

```python
cloudwatch = boto3.client('cloudwatch', region_name=region_
name)
def increment_frame_count(camera_name, count=1):
    cloudwatch.put_metric_data(
        Namespace='VideoAnalysis',
        MetricData=[
        {
            "MetricName": 'FrameCount',
            "Value":count,
            "Unit": 'Count',
            "Dimensions": [
                {
                    "Name": 'camera_name',
                    "Value": camera_name
                },
            ]
        }]
    )
```

Let's generate some random data over 15 minutes to see the results. You can speed up this process by specifying the MetricData's `Timestamp` property instead of relying on the current system time:

```python
from random import randint
for x in range(1,15):
    print("Minute: %s" % x)
    for key in cameras.keys():
        increment_frame_count(cameras[key],
count=randint(1,60))
    sleep(60)
```

Wait for the data generation process to complete, then open the Amazon CloudWatch management console. Navigate to the **All metrics** page and find the `VideoAnalysis` custom namespace.

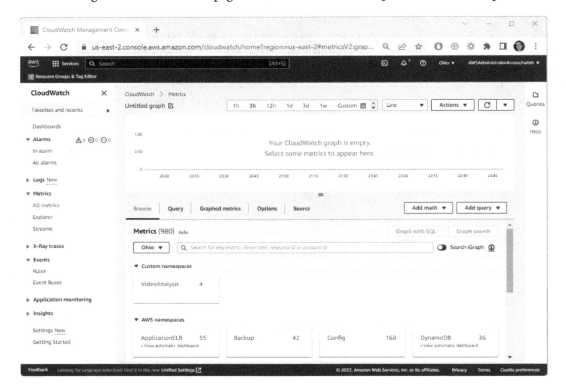

Figure 5.5: Amazon CloudWatch Metrics page

Click into the namespace and add the four metrics. Then, under the **Graphed metrics** tab, set **Statistic** to **Sum** and **Period** to **1 minute**.

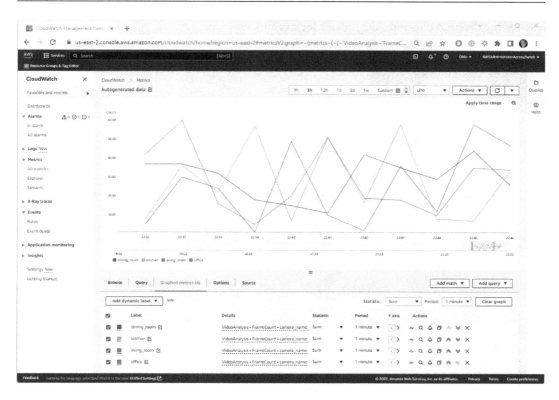

Figure 5.6: Graphed metrics

Quick recap

In this section, you learned how to programmatically connect to IP cameras using the RTSP protocol. After connecting, you could extract and upload individual frames into AWS for further processing.

You can build rich experiences by sampling frames and posting them into the cloud. But what if you want to use the Amazon Rekogition Video APIs? In that case, you upload entire video clips and asynchronously process them. Let's take a look at how that works!

Using the PersonTracking API

Amazon Rekognition Video can track the paths of people in a video stored in an Amazon S3 bucket. You initiate this operation using the StartPersonTracking API, which requires the location of the S3 object and notification channel.

The following diagram illustrates the process for a production system.

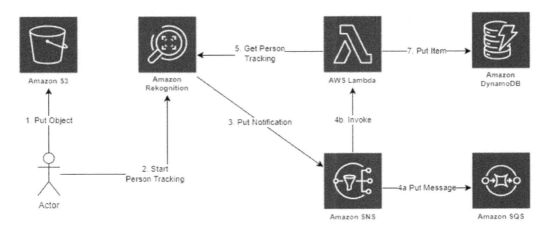

Figure 5.7: The person-tracking process

1. An actor uploads the video into an Amazon S3 bucket.

2. An actor calls the StartPersonTracking API.

3. Amazon Rekognition notifies an Amazon SNS topic of progress.

4. The Amazon SNS topic forwards the notification:

    A. It invokes an AWS Lambda function to handle the callback.

    B. It copies the message for offline troubleshooting.

5. An AWS Lambda function fetches the results using the GetPersonTracking API.

Uploading the video to Amazon S3

This chapter's repository includes a couple of test videos under the tracking folder. Upload those files into the Amazon S3 bucket if not already complete:

```
aws s3 sync 05_VideoAnalysis/tracking s3://ch05-video-use2/
tracking
```

Using the StartPersonTracking API

Amazon Rekognition Video can track the paths of people in a video stored in an Amazon S3 bucket. You initiate this operation using the StartPersonTracking API, which requires the location of the video in S3. Production applications should also specify `NotificationChannel` to receive a notification when the analysis is complete.

You'll need to make the `PUBLISHER_ROLE_ARN` and `TOPIC_ARN` variables equal to the values from the *Technical requirements* section:

```
PUBLISHER_ROLE_ARN='arn:aws:iam::accountid:role/
PersonTrackingPublisher'

TOPIC_ARN='arn:aws:sns:region:accountid:AmazonRekognitionPerson
TrackingTopic'

job = rekognition.start_person_tracking(
    JobTag='HelloFromPackt',
    NotificationChannel={
        'RoleArn': PUBLISHER_ROLE_ARN,
        'SNSTopicArn': TOPIC_ARN
    },
    Video={
        'S3Object':{
            'Bucket': bucket,
            'Name': 'tracking/path_tracking_backyard.MOV'
        }
    })
```

The `StartPersonTracking` API accepts the following parameters.

Parameter Name	Type	Description
ClientRequestToken	String	An idempotent token for preventing the same job from starting multiple times
JobTag	String	An opaque identifier for grouping related jobs in the completion notification
NotificationChannel	NotificationChannel	The Amazon SNS topic to receive the completion notification
Video	Video	The stored video to analyze

Table 5.1: StartPersonTracking request parameters

Next, you can inspect the response using the following snippet:

```
from json import dumps
print(dumps(job, indent=2))
```

Running this command will return a JSON-encoded output similar to the following example. Record the `JobId` value for fetching the results from subsequent calls to the `GetPersonTracking` API. The remaining properties exist for troubleshooting the HTTP client and can be safely ignored:

```
{
    "JobId": "2d14c60409d0360a83bdf33386a9e537698a922456a66aa8e
d0fa67b552f461d",
    "ResponseMetadata": {
        "RequestId": "5acf44e9-79dc-4c0c-b180-de82f067c4f1",
        "HTTPStatusCode": 200,
        "HTTPHeaders": {
            "x-amzn-requestid": "5acf44e9-79dc-4c0c-b180-
de82f067c4f1",
            "content-type": "application/x-amz-json-1.1",
            "content-length": "76",
            "date": "Mon, 26 Dec 2022 19:10:36 GMT"
        },
        "RetryAttempts": 0
    }
}
```

Receiving the completion notification

Amazon Rekognition Video will post a notification to the `PersonTracking` topic when the job completes. You attached an Amazon SQS queue to the topic during the *Technical requirements* section to capture this message. To view its contents, do the following:

1.   Open the Amazon SQS management console.
2.   Select `PersonTrackingQueue` from the queue list.
3.   Select the **Send and receive messages** button.
4.   Select the **Poll for messages** button.

The completion notification message will appear in the **Receive messages** pane within seconds. Select the message's identifier to view the **Message body** contents:

```
{
    "Type" : "Notification",
    "MessageId" : "764410f7-f6f0-599a-ac21-4cd409b8e797",
    "TopicArn" : "arn:aws:sns:us-east-2:581361757134:AmazonReko
gnitionPersonTrackingTopic",
    "Message" :
"{\"JobId\":\"c32127418b81df21df8c800f68848837f7
e21813a4cd33888d07bdb65d611842\",\"Status\":\"SUCCEEDED\",
\"API\":\"StartPersonTracking\",\"Timestamp\":1672084744590,
\"Video\":{\"S3ObjectName\":\"tracking/path_tracking_backyard.
MOV\",\"S3Bucket\":\"ch05-video-use2\"}}",
    "Timestamp" : "2022-12-26T19:59:04.670Z",
    "SignatureVersion" : "1",
    "Signature" : "JVlHD/fAFtzHlKtpW/Yzi3ZLN3L5rAb3E+fb0+
E2WNhWUeEvWYJfbOde/eDW52VsVfykAtaFX2LKM2jcXa3r1Y4Vv+DSzIi0AkKO0
jtA43HvDzSopEZSkYEesPQYRlwrtUsMcGWLNjt/d9QrOLKtXxuGjECDh/3aubKL
WYjzWCR/b4eE+RuvJpfKLqqpcJQxRENMN8bVcEpOoDbkSxcg6M8usg8HXBAMK65
dCqdw3D0T4+kx/vbvJmPODOpnMSmywcSh3D06MriT5or9fJnC/WRxbQfSW
sLlGT3TZfkYaVaPHKj3qrHiynrTKpOokdbPURw/S1d3bBd77ZktSshpQw==",
    "SigningCertURL" : "https://sns.us-east-2.amazonaws.com/
SimpleNotificationService-56e67fcb41f6fec09b0196692625d385.
pem",
    "UnsubscribeURL" : "https://sns.us-east-2.amazonaws.co
m/?Action=Unsubscribe&SubscriptionArn=arn:aws:sns:us-east-
2:581361757134:AmazonRekognitionPersonTrackingTopic:04877b15-
7c19-4ce5-b958-969c5b9a1ecb"
}
```

Suppose the management console doesn't return any messages. This situation is likely due to the job hasn't completed. Use the GetPersonTracking API to retrieve the JobStatus property and confirm:

```
aws rekognition get-person-tracking --job-id
c32127418b81df21df8c800f68848837f7e21813a4cd33888d07bdb65
d611842 | more
{
    "JobStatus": "IN_PROGRESS",
```

```
    "Persons": []
}
```

The `JobStatus` property updates to `SUCCEEDED` or `FAILED` when the job completes. If the message still hasn't appeared, try configuring the access policy on the queue using the `PersonTrackingQueuePolicy.json` file in this chapter's repository:

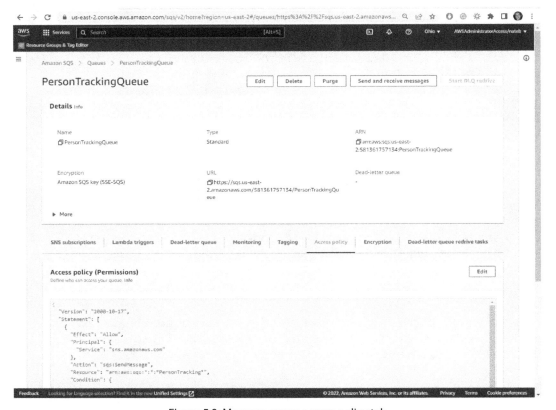

Figure 5.8: Message queue access policy tab

Using the GetPersonTracking API

The completion notification's `Message` property contains the following payload. It states that the `StartPersonTracking` API was successful for the `path_tracking_backyard.mov` file:

```
{
    "JobId": "c32127418b81df21df8c800f68848837f7e21813a4cd33888
d07bdb65d611842",
    "Status": "SUCCEEDED",
```

```
    "API": "StartPersonTracking",
    "Timestamp": 1672084744590,
    "Video": {
        "S3ObjectName": "tracking/path_tracking_backyard.mov",
        "S3Bucket": "ch05-video-use2"
    }
}
```

Use the `JobId` property to fetch the analysis results using the AWS CLI v2 or AWS SDK:

Aws rekognition get-person-tracking –job-id c32127418 b81df21df8c800f68848837f7e21813a4cd33888d07bdb65d611842

Downloading the tracking results from the SDK requires paginating and combining multiple responses. The `get_tracking_results` function implements this pattern by looping until the job completes and traversing through the next set of persons:

```
From time import sleep

def get_tracking_results(jobid):
    combined_response = None
    next_token = None
    while True:
        response = rekognition.get_person_tracking(JobId=jobid)
        if response["JobStatus"] == "IN_PROGRESS":
            print("Job %s is still running..."% jobid)
            sleep(1)
            continue
        if response['JobStaus'] == 'FAILED':
            print('Job %s is failed due to %s' %(
                jobid,
                response['StatusMessage']
            ))
            return None
        if response['JobStatus'] == 'SUCCEEDED':
            combined_response = response
            next_token = response.get('NextToken',None)
            break
```

```
while next_token is not None:
    response = rekognition.get_person_tracking(
        JobId=jobid,
        NextToken=next_token)

combined_response['Persons'].extend(
        response['Persons'])
    next_token = response.get('NextToken',None)

return combined_response
```

Reviewing the GetPersonTracking response

The response has top-level properties for checking the job completion status, error details, and video metadata. Before inspecting the Persons array, you use this information to determine whether the results are usable:

Property Name	Type	Description
JobStatus	String	The status of the job is IN_PROGRESS, SUCCEEDED, or FAILED
NextToken	String	Represents a continuation marker to paginate results
Persons	Array of PersonDetection	The people detected in the video and their tracked path throughout the video
StatusMessage	String	Descriptive error message for FAILED jobs
VideoMetadata	VideoMetadata	Information about the video that Amazon Rekognition Video analyzed

Table 5.2: GetPersonTracking response

The PersonDetection object contains details about an individual and their position at various time offsets. This structure includes top-level properties of PersonDetail and the video's relative timestamp (milliseconds).

PersonDetail objects won't always have every property populated. For instance, if the person is facing away from the camera, the Face information isn't available for that frame.

It's also important to note that the index can exceed the total number of people in the video clip. For example, `path_tracking_backyard.mov` detects 10 people despite there being one person. This disconnect becomes more pronounced in action sequences and videos that film people from multiple angles.

The following table enumerates the `FaceDetail` object properties:

Property Name	Type	Description
BoundingBox	BoundingBox	A bounding box around the detected person
Face	FaceDetail	Face details for the detected person
Index	Long	An identifier of the person detected within a video

Table 5.3: PersonDetail properties

This particular video's analysis is over 2,500 lines in length. The complete response is available as `GetPersonTrackingResponse.json` in the chapter's repository.

Viewing the frame

Now for our final magic trick. Let's render the `BoundingBox` instance specified in a `PersonDetection` instance. This task requires the following steps:

1. Use OpenCV to open the video.
2. Jump to the specified `Timestamp` offset.
3. Calculate the `BoundingBox` rectangle.
4. Draw the `BoundingBox` rectangle.
5. Persist and view the drawing.

First, you must download the video from Amazon S3 to the local filesystem so the `VideoCapture` class can open and interact with it:

```
from cv2 import VideoCapture
capture = VideoCapture('tracking/path_tracking_backyard.mov')
```

> **Handling large files**
>
> Suppose the videos are too massive for downloading, or your application must fetch them multiple times. You can move them from Amazon S3 into Amazon **Elastic File System** (**EFS**) or Amazon FSx for Windows. OpenCV supports directly reading from these network filesystems after mounting them.
>
> See `https://docs.aws.amazon.com/efs/latest/ug/mounting-fs.html` and `https://docs.aws.amazon.com/fsx/latest/WindowsGuide/using-file-shares.html` for more information.

Next, let's select the first `PersonDetection` object that's forward facing. You can find this result by enumerating the `Persons` property and checking for the presence of a `Face` child object:

```
tracking_results = get_tracking_results(job_id)
first_detection = None
for detection in tracking_results['Persons']:
    if 'Face' in detection['Person']:
        first_detection = detection
        break
```

After finding this instance, you can `set` the `VideoCapture` object to that frame's `Timestamp` offset. Then, use the `read` function to receive a tuple representing the `result` and `frame`.

If the operation is successful, `result` will equal `True`, and the frame will contain an array of pixel values:

```
from cv2 import CAP_PROP_POS_MSEC
capture.set(
    CAP_PROP_POS_MSEC,
    first_detection['Timestamp'])
result, frame = capture.read()
```

You can safely release (using `release`) the `capture` object and free resources maintained by OpenCV. Forgetting to call this method will result in a memory leak and eventually crash the process:

```
capture.release()
print('The capture variable is %s' % capture.isOpened())
```

Third, you'll need to denormalize `BoundingBox` and calculate its rectangle. This step should feel reminiscent of the skateboard example in *Chapter 2*:

```
def denormalize_bounding_box(bounding_box, image_size):
    width = int(bounding_box['Width'] * image_size[0])
    left = int(bounding_box['Left'] * image_size[0])

    height = int(bounding_box['Height'] * image_size[1])
    top = int(bounding_box['Top'] * image_size[1])
    return (top,left,height,width)


video_metadata = tracking_results['VideoMetadata']
image_size = (
  video_metadata['FrameWidth'],
  video_metadata['FrameHeight']
)


(top,left,height,width) = denormalize_bounding_box(
    first_person['Person']['BoundingBox'],
    image_size)
```

Fourth, convert the `frame` instance into a PIL `Image` and draw a red rectangle around the person. Specifying the rectangle's `width` value controls the line's thickness:

```
from PIL import Image, ImageDraw
image = Image.fromarray(frame)

drawing = ImageDraw.Draw(image)
drawing.rectangle(
    xy=(left,top,left+width, left+height),
    outline='red',
    width=5)
```

Lastly, use the `save` or `show` function to render `image`:

```
image.save('tracking/ViewFrame.jpeg',format='JPEG')
image.show()
```

This chapter's repository contains the final `ViewFrame.jpeg` rendering. It is also shown here for reference.

Figure 5.9: ViewFrame.jpeg

Quick recap

In this section, you learned how to use the Amazon Rekognition Video API to track a person's movements through a video clip. You also configured a serverless event notification channel, which avoids needing to pull results continuously. This pattern will reappear multiple times across the AWS AI services because it's incredibly efficient at any scale.

Summary

You have successfully built a video analysis pipeline for real-time video streams using OpenCV and the RTSP protocol to sample frames from IP cameras. You processed a short recording from my iPhone using the Amazon Rekognition Video API to identify time offsets and PIL to draw bounding boxes around people.

You assembled an (almost) production-ready pipeline that asynchronously processes files and uses a completion notification to avoid pulling for status. This foundational API opens the door to sports highlights, safety systems, and improving building layouts, among other potential scenarios. Suppose a traditional store reviewed the footage and determined that its customers gravitate toward specific areas. In that case, they should place high-margin items in those areas to increase sales.

One of the limitations of our solution is that indexes, not names, reference humans. In the last chapter, you mitigated this issue using Amazon Rekognition's Face API to build a contactless casino and resort. Just think of the many situations that could combine these two chapters' material!

In the next chapter, you'll learn how to build safe, inclusive, online communities with oceans of user-generated content. Amazon Rekognition and other AWS AI services make it easy to inspect third-party artifacts and programmatically enforce custom guidelines for your users. When you're ready, let's continue onward to content moderation!

6
Moderating Content with AWS AI Services

Modern web and mobile applications need capabilities for users to collaborate and socialize. These features are becoming table stakes, with over 80% of the internet becoming user-generated content. Think about the sheer volume of online interactions, product reviews, and brand interactions you've made this week alone. It's no wonder they call it the user-generated content era.

In parallel, mobile devices are evolving how customers expect to engage with social features. They demand multi-modal capabilities that span audio, video, and rich-text documents. While this drives platform adoption, it also attracts trolls that bring toxicity and extremism.

Traditional businesses ask their customers and human moderators to flag inappropriate content. However, relying on customers to report negative experiences isn't a good strategy and will hurt engagement. Meanwhile, maintaining a human workforce is too expensive to scale. Building custom automation that looks for inappropriate content across these media types is a huge undertaking.

AWS AI services enable building content moderation workflows that work across mainstream media types to enforce your business policies. You can leverage these capabilities without machine learning expertise using simple APIs.

In this chapter, you will learn how to do the following:

- Use the `DetectContentModeration` API for images
- Build automation to react to detections
- Use the `DetectContentModeration` API for videos
- Track content moderation detections using Amazon CloudWatch

Technical requirements

A Jupyter notebook is available to run the example code from this chapter. You can access the most recent code from this book's GitHub repository: https://github.com/PacktPublishing/Computer-Vision-on-AWS. Clone that repository to your local machine using the following command:

```
$ git clone https://github.com/PacktPublishing/Computer-Vision-
on-AWS
$ cd Computer-Vision-on-AWS/06_ContentModeration
```

Additionally, you will need an AWS account and the Jupyter Notebook. *Chapter 1*, contains detailed instructions for configuring the developer environment.

Moderating images

Suppose you're adding social features to a swimming club's website. The club wants members to post their user-generated content but to avoid alcohol associations. It also allows content featuring appropriate swimwear attire provided it's not sexualized.

Amazon Rekognition's DetectModerationLabels API can identify unsafe content within JPEG and PNG format images. This action lets you quickly implement categorical filters such as nudity, extremism, and visually disturbing content.

Amazon Rekognition handles this scenario through its top-level and secondary-level labeling schemes (see *Table 6.2*). You can implement coarse and granular rules to match the swimming club's requirements.

Using the DetectModerationLabels API

Amazon Rekognition's DetectModerationLabels API can identify unsafe content within JPEG and PNG format images. This action lets you quickly implement categorical filters such as nudity, extremism, and visually disturbing content.

This snippet demonstrates how to invoke the DetectModerationLabels API for a given image. Like the previous chapter's examples, it creates the Amazon Rekognition client and defaults the AWS Region to us-east-2. The main function represents the script's entry point, which invokes the DetectModerationLabels API and prints the response labels. See *Chapter 2*, for more information:

```
import boto3

bucket_name = 'cv-on-aws-book-nbachmei'
photo='06_ContentModeration/images/swimwear.jpg'
region_name = 'us-east-2'
rekognition = boto3.client('rekognition', region_name=region_
```

```
name)

def moderate_image(photo, bucket):
    response = rekognition.detect_moderation_
labels(Image={'S3Object':{'Bucket':bucket,'Name':photo}})

    print('Detected labels for ' + photo)
    for label in response['ModerationLabels']:
        print (label['Name'] + ' : ' +
str(label['Confidence']))
        print (label['ParentName'])
    return len(response['ModerationLabels'])

def main():
    label_count=moderate_image(photo, bucket_name)
    print("Labels detected: " + str(label_count))
```

When you call the DetectModerationLabels API, there are several optional parameters available. This snippet represents the current request payload:

```
{
    "HumanLoopConfig": {
        "DataAttributes": {
            "ContentClassifiers": [ "string" ]
        },
        "FlowDefinitionArn": "string",
        "HumanLoopName": "string"
    },
    "Image": {
        "Bytes": blob,
        "S3Object": {
            "Bucket": "string",
            "Name": "string",
            "Version": "string"
        }
    },
    "MinConfidence": number
}
```

The request envelope contains properties we've seen in previous chapters, such as the Image object and the MinConfidence threshold for specifying the input image and accuracy requirements.

On the other hand, HumanLoopConfig represents the integration point with **Amazon Augmented AI (Amazon A2I)**. Amazon A2I allows you to incorporate human review into ML applications based on specific requirements. For instance, your workload might require additional approval for situations that lack confidence in a purely automated solution.

We'll explore Amazon A2I in *Chapter 11*, Until then, it's sufficient to know the service exists and can bring human moderators into your DetectModerationLabel API calls.

The following table enumerates the DetectModerationLabel API's parameters.

Property	Required	Type	Description
HumanLoopConfig	No	HumanLoopConfig object	Sets up the configuration for human evaluation
Image	Yes	Image object	The input image as base64-encoded bytes or an Amazon S3 object
MinConfidence	Default (50)	Float between 0 and 100	Specifies the minimum confidence level for the labels to return

Table 6.1: Detect moderation label request

Using top-level categories

First, let's automate checking for alcohol within images. This function accepts the response from the DetectModerationLabel API. Next, the code iterates through the ModerationLabels list and inspects the top_level category:

```
def contains_alcohol(detect_moderation_response):
    for label in detect_moderation_
response['ModerationLabels']:
        top_level = label['ParentName']
        confidence = label['Confidence']

        if top_level == "Alcohol":
            print('Alcohol detected - %2.2f%%.' % confidence)
            return True
```

```
    return False
```

Using secondary-level categories

The second design requirement asks that we permit swimwear but avoid sexualized content. For this situation, you must combine top-level and secondary-level category filters:

```
ALLOWED_SUGGESTIVE_LABELS= [
    'Female Swimwear Or Underwear',
    'Male Swimwear Or Underwear']


def contains_appropriate_attire(detect_moderation_response):
    for label in detect_moderation_
response['ModerationLabels']:
        top_level = label['ParentName']
        secondary_level = label['Name']
        confidence = label['Confidence']

        if top_level == "Explicit Nudity":
            print('Explicit Nudity detected - %2.2f%%.' %
confidence)
            return False

        if top_level == "Suggestive":
            if secondary_level not in ALLOWED_SUGGESTIVE_
LABELS:
                print('Prohibited Suggestive[%s] detected -
%2.2f%%.' %
                (secondary_level, confidence))
                return False
            else:
                print('Allowed Suggestive[%s] detected -
%2.2f%%.' %
                (secondary_level, confidence))

    return True
```

Putting it together

This section aims to build a content moderation filter for a swimming club's website. Now that we have all the pieces, let's put them all together.

Here, you'll notice the `moderate_image` function brings our rules together in an almost English expression. The `ModerateImages.py` file in this chapter's repository combines these snippets into a complete script:

```
def moderate_image(photo, bucket):
    response = rekognition.detect_moderation_labels(
        Image={
            'S3Object':{
                'Bucket': bucket,
                'Name':photo
            }
        })

    if not contains_alcohol(response) and contains_appropriate_
attire(response):
        return True
    return False
```

The following table enumerates the `DetectModerationLabel` categories. While you cannot directly extend these options, AWS does release periodic updates every 3 to 6 months.

Top-Level Category	Second-Level Category
Explicit Nudity	Nudity
	Graphic Male Nudity
	Graphic Female Nudity
	Sexual Activity
	Illustrated Explicit Nudity
	Adult Toys

Top-Level Category	Second-Level Category
Suggestive	Female Swimwear Or Underwear
	Male Swimwear Or Underwear
	Partial Nudity
	Barechested Male
	Revealing Clothes
	Sexual Situations
Violence	Graphic Violence Or Gore
	Physical Violence
	Weapon Violence
	Weapons
	Self Injury
Visually Disturbing	Emaciated Bodies
	Corpses
	Hanging
	Air Crash
	Explosions And Blasts
Rude Gestures	Middle Finger
Drugs	Drug Products
	Drug Use
	Pills
	Drug Paraphernalia
Tobacco	Tobacco Products
	Smoking
Alcohol	Drinking
	Alcoholic Beverages
Gambling	Gambling

Top-Level Category	Second-Level Category
Hate Symbols	Nazi Party
	White Supremacy
	Extremist

Table 6.2: Detect moderation labels

Quick recap

In this section, you learned about the `DetectModerationLabels` API and how it helps protect users from toxicity and inappropriate content. You also learned how to combine different classifications for granular and fine-grained policies. Next, you'll explore applying these capabilities to stored video files.

Moderating videos

Collaborating using plain text and images has become so 2000 and late. Nearly everyone has a mobile device with high-speed connectivity, and this is evolving the user-generated content era to video formats. Amazon Rekognition supports video content moderation using its `StartContentModeration` and `GetContentModeration` APIs.

Amazon Rekognition can start the asynchronous detection of inappropriate, unwanted, or offensive content in a stored video. To access this capability, you need to take the following actions:

1. Upload the video into an Amazon S3 bucket.

2. Call the `StartContentModeration` API.

3. Amazon Rekognition Video will begin processing the video.

4. Amazon Rekognition Video posts notifications to a topic.

5. You subscribe to that topic using Amazon SQS and AWS Lambda.

This architecture should seem familiar after the *PersonTracking* section of *Chapter 5*, Amazon Rekognition Video posts a completion notification in both use cases that triggers your custom code.

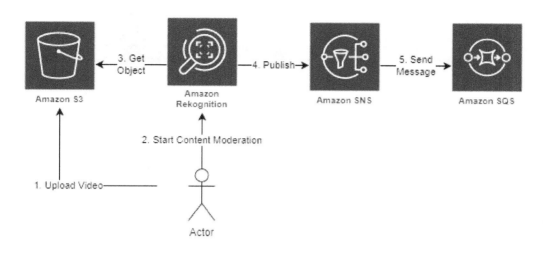

Figure 6.1: Video moderation architecture

Creating the supporting resources

This chapter's repository contains an AWS CloudFormation template for creating the resources needed in this section. To provision an AWS CloudFormation template, you need to do the following:

1. Navigate to `https://us-east-2.console.aws.amazon.com/cloudformation/home?region=us-east-2`.

2. Activate the **Create stack** dropdown and choose **With new resources (standard)**.

3. In the **Specify template** section, set **Template source** to **Upload a template file**.

4. Select the **Choose file** button and specify the repository file `VideoContentModeration.template.json`.

5. Select the **Next** button.

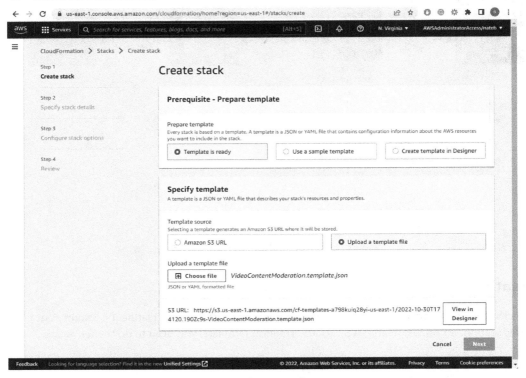

Figure 6.2: Create stack page

6. Set **Stack name** to `VideoContentModeration`.

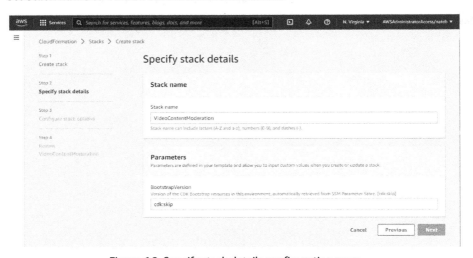

Figure 6.3: Specify stack details configuration page

7. Leave all defaults on the **Configure stack options** page.

8. Select the **Next** button.

9. Scroll to the bottom of the **Review VideoContentModeration** page.

10. Check the **I acknowledge that AWS CloudFormation might create IAM resources with custom names** checkbox.

11. Select the **Submit** button.

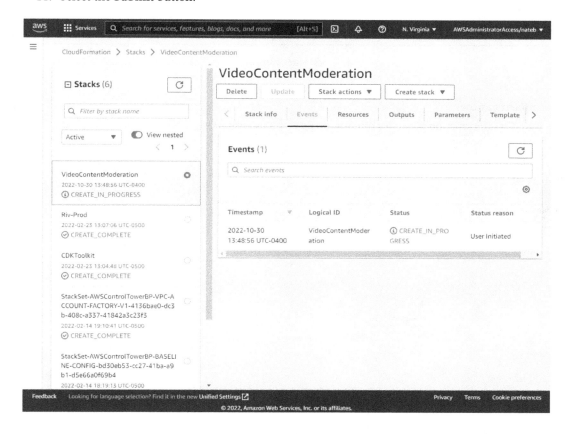

Figure 6.4: Wait for the deployment screen

Finding the resource ARNs

AWS CloudFormation will take 2 to 3 minutes to provision the template. The console periodically refreshes and will eventually set **Status** to CREATE_COMPLETE. Next, navigate to the **Outputs** tab and note the values there.

The following table enumerates the exported AWS CloudFormation outputs and their purpose:

Export-Name	Description
NotificationChannel-RoleArn	The role used by Amazon Rekognition for publishing updates
NotificationChannel-SNSTopicArn	The topic for receiving Amazon Rekognition notifications
Video-InputBucket	The input Amazon S3 bucket for holding video files

Table 6.3: Stack outputs

> **Important note**
> AWS CloudFormation generates random `suffixes` on provisioned resources, so your specific values will be slightly different.

This is what it looks like:

Figure 6.5: Output for VideoContentModeration

Uploading the sample video to Amazon S3

This chapter's repository includes the one-minute `contentmoderationsample.mp4` video. It aims to demonstrate content moderation while simultaneously being all-age appropriate through imagery containing suggestiveness and alcohol. The internet is full of toxic garbage, so using your preferred search engine to find examples of more vulgar content won't take much effort.

Use the **AWS Command Line Interface** (**AWS CLI v2**) to upload this file into the Video-InputBucket (from the AWS CloudFormation stack's outputs):

```
$ aws s3 cp contentmoderationsample.mp4 s3://
videocontentmoderation-bucket-****
```

Using the StartContentModeration API

After deploying all dependent resources, you can begin experimenting with the StartContentModeration API. Create a new Juypter notebook and specify your configuration information in the first code block. Replace the **** placeholders with the values from the AWS CloudFormation stack outputs:

```
import boto3
region_name = 'us-east-2'
video_input_bucket = 'videocontentmoderation-bucket-****'
notification_channel_rolearn = 'arn:aws:iam::ACCOUNTID:role/
VideoContentModeration-RekognitionRole****'
notification_channel_snstopic_arn = 'arn:aws:sns:REGION:
ACCOUNTID:VideoContentModeration-Topic****'
rekognition = boto3.client(
   'rekognition',
   region_name=region_name)
```

Using the boto3 client, invoke the StartContentModeration API. You'll need to specify the notification channel and Amazon S3 video location. When Amazon Rekognition Video finishes processing the file, it will publish a message to the selected topic:

```
moderation_job = rekognition.start_content_moderation(
    NotificationChannel={
        'RoleArn': notification_channel_rolearn,
        'SNSTopicArn': notification_channel_snstopic_arn
    },
    Video={
        'S3Object':{
            'Bucket':video_input_bucket,
            'Name':'contentmoderationsample.mp4'
        }
    }
)
```

> **Important note**
>
> If you receive an `InvalidS3ObjectException` error, confirm the configuration values of the `region_name` and `video_input_bucket` variables.
>
> `InvalidS3ObjectException: An error occurred (InvalidS3ObjectException) when calling the StartContentModeration operation: Unable to get object metadata from S3. Check object key, region, and/or access permissions.`

The following table enumerates the `StartContentModeration` API parameters. Specifying the `NotificationChannel` is optional but strongly recommended for production workloads. Using this mechanism avoids continuously pulling the job's status, helping to reduce costs.

Suppose you specified the optional `ClientRequestToken` property. In that case, Amazon Rekognition will return the same JobId instead of creating new starting new jobs. This property is a convenient mechanism for your retry code to avoid accidentally processing and paying for the same file multiple times.

Property	Required	Type	Description
ClientRequestToken	No	String	The idempotent token prevents the same job from accidentally starting more than once.
JobTag	No	String	An identifier you specify that's returned in the completion notification.
MinConfidence	No	Float	Specifies the confidence threshold for returning a moderated content label.
NotificationChannel	No	NotificationChannel	The Amazon SNS topic that you want Amazon Rekognition Video to publish completion notifications for.
Video	Yes	Video	The video in which you want to detect inappropriate, unwanted, or offensive content.

Table 6.4: Start content moderation parameters

The `moderation_job` response contains the `JobId` property, the primary identifier for tracking this operation:

```
{
    "JobId": "deadbeef00000042c6c00dcc6c9b60669e441926136",
}
```

Examining the completion notification

Amazon Rekognition will process the example file within a few moments and publish a notification to the NotificationChannel-SNSTopicArn. The AWS CloudFormation template subscribes an **Amazon Simple Queuing Service (Amazon SQS)** queue to the topic for troubleshooting purposes. To retrieve these messages, do the following:

1. Navigate to https://us-east-2.console.aws.amazon.com/sqs/v2/home?region=us-east-2.

2. Choose the VideoContentModeration-AuditQueue queue.

3. Select the **Send and receive messages** button.

4. Select the **Poll for messages** button.

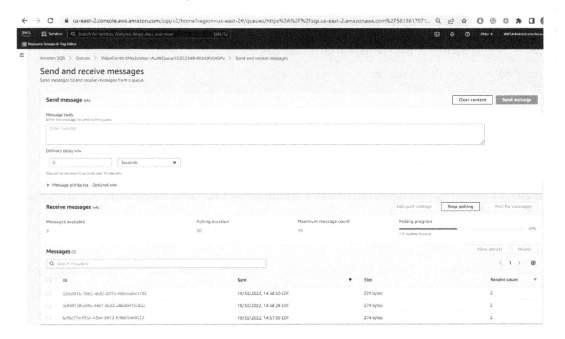

Figure 6.5: Amazon SQS Send and receive messages console

Selecting a message identifier will display the notification's payload:

```
{
    "JobId": "deadbeef00000042c6c00dcc6c9b60669e441926136",
    "Status": "SUCCEEDED",
    "API": "StartContentModeration",
    "Timestamp": 1667156330321,
    "Video": {
        "S3ObjectName": "contentmoderationsample.mp4",
        "S3Bucket": "videocontentmoderation-bucket-****"
    }
}
```

Property	Type	Description
JobId	String	The identifier for this stored video operation
Status	Enum	A flag indicating whether the job has a status of SUCCEEDED, FAILED, or IN_PROGRESS
API	String	The completed operation name
Timestamp	Epoch	Unix timestamp of the notification
Video	Video	The location of the analyzed stored video file

Table 6.5: Amazon Rekognition Video notification

Using the GetContentModeration API

You can use the GetContentModeration API to query the **Status** and fetch results from the StartContentModeration job. The GetContentModeration API is a paginated operation and only returns up to 1,000 results. Beyond this limit, you must specify the NextToken marker to signify subsequent pages.

The get_complete_content_moderation_results utility function will paginate the results and combine them into one logical response structure:

```python
def get_complete_content_moderation_results(jobId):
    next_token = None
    complete_response = None
    while True:
        if next_token:
```

```
        response = rekognition.get_content_moderation(
            JobId=jobId,
            NextToken=next_token)
        complete_response['ModerationLabels'].
extend(response['ModerationLabels'])
        else:
            response = rekognition.get_content_moderation(
                JobId=jobId)
            complete_response = response

        if 'NextToken' in response:
            next_token = response['NextToken']
        else:
            break
    return complete_response


combined_response =
 get_complete_content_moderation_results(moderation_
job['JobId'])
```

Next, examine the `get_complete_content_moderation_results` function's combined response. There are over 200 moderation labels, so only an excerpt is shown here. You can find the complete response in the `GetContentModerationVideoResponse.json` file within the chapter's repository.

The following table enumerates the `GetContentModeration` response's properties. We have already used several properties to monitor and fetch the analysis results. So, let's focus our attention on the `VideoMetadata` and `ModerationLabels` properties.

Property	Type	Description
JobStatus	Enum	The content moderation analysis job's Status
StatusMessage	String	Descriptive error for failed jobs
VideoMetadata	VideoMetadata	Information about a video that was analyzed by Amazon Rekognition

ModerationLabels	Array of ContentModerationDetection	The detected inappropriate, unwanted, or offensive content moderation labels and their detected time(s)
NextToken	String	The pagination token to use while fetching the moderation labels
ModerationModelVersion	String	The moderation detection model's version number

Table 6.6: GetContentModeration response

The `VideoMetadata` property provides information about the video encoding, total runtime, and the frame's dimensions. For example, this QuickTime video is 60 seconds long with 30 frames per second at a width of 1080x1920 width:

```
"VideoMetadata": {
    "Codec": "h264",
    "DurationMillis": 60066,
    "Format": "QuickTime / MOV",
    "FrameRate": 30.0,
    "FrameHeight": 1080,
    "FrameWidth": 1920,
    "ColorRange": "LIMITED"
}
```

The `ModerationLabels` property contains an array of `ContentModerationDetection` objects. Each detection represents an inappropriate, unwanted, or offensive instance within the video. Amazon Rekognition Video uses a frame sampling procedure, so the reported `Timestamp` offset approximates when `ModerationLabel` first appears:

```
{
    "Timestamp": 2000,
    "ModerationLabel": {
      "Confidence": 90.82994079589844,
      "Name": "Revealing Clothes",
      "ParentName": "Suggestive"
    }
}
```

Quick recap

In this section, you learned how to use the `StartContentModeration` and `GetContentModeration` APIs to detect unwanted and inappropriate content within stored video files. You also learned how to configure event-driven code that's highly scalable and economical. Now let's make these ideas more tangible in code as AWS Lambda functions.

Using AWS Lambda to automate the workflow

AWS Lambda is a serverless, event-driven computing service that lets you run code for virtually any application or backend service (that's Linux-compatible). Under the covers, it uses a purpose-built operating system called Firecracker to scale your code elastically within milliseconds. This capability allows Lambda to support burst traffic patterns and AWS to charge per millisecond that your code is running!

There are three phases to a Lambda function's life cycle: `Init`, `Invoke`, and `Shutdown`. During the `Init` phase, your code should fetch configuration information and prepare any stateless clients. Next, the Lambda service will invoke the `lambda_handler` entry point one or more times. After processing all pending events, the service eventually releases the underlying Firecracker micro-virtual machine.

Now that you've had a 30-second crash course on the service, let's use it to operationalize the entire video moderation content workflow in AWS.

The following architecture depicts our target state, reusing several components from the previous *Moderating videos* section.

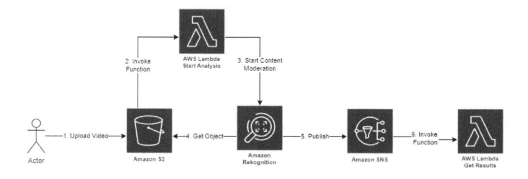

Figure 6.5: Video Moderation Architecture

After introducing the Lambda functions, the event-driven sequences will be the following:

1. A user uploads a video file into the `VideoContentModerationBucket`.
2. The incoming file triggers the `StartAnalysis` function.
3. The `StartAnalysis` function calls the `StartContentModeration` API.
4. Amazon Rekognition Video will process the video file.
5. Amazon Rekognition Video will `Publish` the completion notification to an Amazon SNS topic.
6. Amazon SNS will invoke the `GetAnalysis` function.

Since this chapter's AWS CloudFormation template has already created the resources, you can focus on implementing the logic for *steps 3* and *6*. You can find completed AWS Lambda functions in the chapter's repository under the `video` folder.

Implement the Start Analysis Handler

You'll need to create a new file called `start-analysis_function.py` to receive the incoming file notification from the `Video-InputBucket`. This function's code must complete the following subtasks:

1. Initialize the AWS Lambda function's configuration.
2. Confirm the incoming file is supported.
3. Invoke the `StartContentModeration` API.
4. Define the `lambda_handler` entry point.

First, declaring a Python function's `Initialization` phase is relatively straightforward: we merely import any modules and declare any global state like traditional scripts. You can avoid hardcoding resource names by passing them as environment variables and accessing them through the `environ` class:

```
import boto3
from json import dumps
from os import environ
rekognition = boto3.client('rekognition')

NOTIFICATION_CHANNEL_ROLEARN =    environ.get('NOTIFICATION_
CHANNEL_ROLEARN')
NOTIFICATION_CHANNEL_SNSTOPIC_ARN = environ.get('NOTIFICATION_
CHANNEL_SNSTOPIC_ARN')
```

Next, we need to confirm the incoming file is valid. The `StartContentModeration` API presently supports MP4 and MOV files. We'll assume that incoming files with appropriate extensions are compatible.

A more sophisticated application could fetch the file's header by specifying the Range property to the Amazon S3 GetObject API. This approach restricts the download size to only a few kilobytes versus an arbitrarily large file:

```
def is_supported_file(name):
    name = str(name).lower()
    if name.endswith('.mp4') or name.endswith('.mov'):
        return True
    return False
```

Third, let's create a function that accepts an incoming file's location and forwards it to the StartContentModeration API. This function should look very familiar having completed the *Moderating videos* section:

```
def process_file(bucket, name):
    moderation_job = rekognition.start_content_moderation(
        NotificationChannel={
            'RoleArn': NOTIFICATION_CHANNEL_ROLEARN,
            'SNSTopicArn': NOTIFICATION_CHANNEL_SNSTOPIC_ARN
        },
        Video={
            'S3Object':{
                'Bucket':bucket,
                'Name':name
            }
        })
    print(dumps(moderation_job, indent=2))
```

Lastly, define the lambda_handler function entry method once per event. If you rename this function's name, update the Lambda configuration's handler.

When the Lambda service invokes this function, it passes an asynchronous Amazon S3 notification event and the runtime context. We'll iterate through the records and extract the location of the incoming files.

Then, if that location passes the is_supported_file check, it's forwarded to Amazon Rekognition Video using the process_file function:

```
def lambda_handler(event, context):
    for record in event['Records']:
        bucket = record['s3']['bucket']['name']
```

```
        name = record['s3']['bucket']['object']['key']

        if not is_supported_file(name):
            print("UNSUPPORTED FILE: s3://{}/{})".
format(bucket,name))
        else:
            print("StartContentModeration(s3://{}/{})".
format(bucket,name))
            process_file(bucket, name)
```

> **Important note**
> The Amazon S3 bucket only runs this function when the created file has an `.mp4` or `.mov`
> extension (case-sensitive). You can support more file types using AWS Elemental or OpenCV,
> which are beyond this chapter's scope.

Implementing the Get Results Handler

Next, let's create a second file called `get-results_function.py` to contain the Lambda function
for responding to the completion notification. The business logic for this function must do the following:

1. Initialize the Lambda function.

2. Declare the `lambda_handler` entry point.

3. Call the `get_complete_content_moderation_results` utility and summarize the
    results in a frequency distribution.

4. Publish the distribution in Amazon CloudWatch metrics.

First, you declare the initialization similar to the previous `StartAnalysis` function. There's no
configuration necessary, so you only import the required modules and create the boto3 clients:

```
import boto3
from json import loads

rekognition = boto3.client('rekognition')
cloudwatch = boto3.client('cloudwatch')
```

Next, define a `lambda_handler` function to process the Amazon SNS completion notification.
The Lambda service passes this information as the `event` parameter.

The function implementation iterates through the event's `Records` and parses the JSON-encoded `Message` property. See the *Examining the completion notification* section in this chapter for more information.

Pass `record_message` to the `get_moderation_frequency` function to get the distribution frequency. Afterward, use the `publish_metrics` helper function to persist them into Amazon CloudWatch:

```python
def lambda_handler(event,context):
    for record in event['Records']:
        record_message = loads(record['Sns']['Message'])
        print(record_message)
        frequency = get_moderation_frequency(record_message)
        publish_metrics(frequency)
```

Third, use the `get_complete_content_moderation_results` utility function to retrieve the content moderation results from the `GetContentModeration` API. We need to do something with these results, so let's create a histogram of the total times a reported label count:

```python
def get_moderation_frequency(notification):
    moderation_results = get_complete_content_moderation_
results(notification['JobId'])
    parent_frequency = {}
    for label in moderation_results['ModerationLabels']:
        name = label['ModerationLabel']['Name']
        parent = label['ModerationLabel']['ParentName']

        if len(parent) == 0:
            parent = 'TopLevel'
        if len(name) == 0:
            name = "None"

        if parent not in parent_frequency:
            parent_frequency[parent]= { name: 1 }
        else:
            if name not in parent_frequency[parent]:
                parent_frequency[parent][name]=1
            else:
                parent_frequency[parent][name] += 1
    return parent_frequency
```

Finally, define the `publish_metrics` function for writing the frequency information into Amazon CloudWatch metrics. The `cloudwatch` client uses the `PutMetricData` API to publish in the `VideoContentModeration` custom namespace created on first access:

```python
def publish_metrics(frequency):
    metric_data = []
    for (parent,secondary) in frequency.items():
        metric_data.append({
            'MetricName':'ContentModeration',
            'Dimensions': [
                {
                    'Name': 'TopLevel',
                    'Value': str(parent),
                }
            ],
            'Value': sum(secondary.values()),
            'Unit':'Count',
        })
        for (name, count) in secondary.items():
            metric_data.append({
                'MetricName':'ContentModeration',
                'Dimensions': [
                    {
                        'Name': 'TopLevel',
                        'Value': str(parent),
                        'Name': 'Secondary',
                        'Value': str(name),
                    }
                ],
                'Value': count,
                'Unit':'Count',
            })

    cloudwatch.put_metric_data(
        Namespace='VideoContentModeration',
        MetricData=metric_data)
```

Publishing function changes

You can publish this version of the Lambda function using the AWS console:

1. Navigate to `https://us-east-2.console.aws.amazon.com/lambda/home?region=us-east-2#/functions`.

2. Select the existing `VideoContentModeration-StartAnalysisFunction` function.

3. Activate the **Code** tab and double-click on the **Code source** explorer control.

4. Replace the current code with your version.

5. Click the **Deploy** button.

6. Repeat steps 2 through 5 for `VideoContentModeration-GetResultsFunction`.

Experiment with the end-to-end

After you deploy these changes to the Amazon Lambda function, it will write the `Content Moderation` label frequency into Amazon CloudWatch. Go ahead and search for more files and upload them into the `Video-InputBucket`. Amazon S3 will notify your video moderator that new files are available and will automatically call the `Start Content Moderation` API.

Summary

The user-generated content era represents opportunities for users to collaborate and socialize in rich media formats. The growth of user communities in both size and diversity drives engagement around your products and services. For web and mobile platforms to monetize that traffic, they require capabilities to build safe and inclusive environments. These communities will become toxic and have inappropriate content without adequate protection and governance structures.

Amazon Rekognition can moderate images without needing a data science team. You learned how to use the `DetectContentModeration` API to quickly and easily discover top-level and secondary inappropriate categories. Then, you built automation to flag content on granular and coarse levels depending on your business requirements.

These capabilities are critical to modern platforms but didn't go far enough. End users have expectations of content beyond images, in the form of video clips. You learned how to build a real-time event architecture for analyzing video files at scale. You can now upload any file into your Amazon S3 bucket and automatically process it. The results persist to Amazon CloudWatch metrics so that you can monitor trends.

In the next chapter, you'll learn about Amazon Lookout for Vision. Manufacturers use Amazon Lookout for Vision to automate quality inspection and discover defects in parts and assembly processes. Turn the page and continue onward when you're ready.

Part 3: CV at the edge

This third part consists of two cumulative chapters that will provide a detailed overview of Amazon Lookout for Vision and how it applies to CV at the edge.

By the end of this part, you will understand the features of Lookout for Vision, how to train a model to detect anomalies, and the benefits of deploying a model at the edge.

This part comprises the following chapters:

- *Chapter 7, Introducing Amazon Lookout for Computer Vision*
- *Chapter 8, Detecting Manufacturing Defects Using CV at the Edge*

Introducing Amazon Lookout for Vision

So far in this book, we've talked about using **computer vision** (**CV**) to identify labels, objects, or text in images and videos. However, another key application of CV is identifying defects or anomalies in industrial manufacturing processes. Imagine that you're a medical products manufacturer; you will need to deploy dozens of inspectors to control the quality of the products being shipped out. This manual inspection task is expensive, prone to error, and does not scale. Leveraging CV and **artificial intelligence** (**AI**)/**machine learning** (**ML**)-based solutions such as **Amazon Lookout for Vision** allows customers to drive efficiency and enhance the productivity of human operators. Amazon Lookout for Vision allows you to identify defects, such as missing components on circuit boards, damages such as scratches or bent parts, and defects with a repeating pattern, indicating process issues.

Amazon Lookout for Vision is a fully managed service that allows you to build automated quality inspection pipelines without any ML experience. You can use it to perform image classification (whether an image contains an anomaly) or image segmentation (identifying the location of an anomaly).

By the end of this chapter, you should have a better understanding of Amazon Lookout for Vision, and the process to train and deploy models to detect anomalies such as whether a pill is damaged or not. You can similarly create and host models using Amazon Lookout for Vision to identify defects specific to your use case.

This chapter will cover the following topics:

- Introducing Amazon Lookout for Vision
- Creating a model using Lookout for Vision
- Building a model to identify damaged pills

Technical requirements

You will require the following:

- Access to an active **Amazon Web Services** (**AWS**) account with permissions to access Amazon SageMaker and Amazon Rekognition

- PyCharm or any Python IDE

- All the code examples for this chapter can be found on GitHub at `https://github.com/PacktPublishing/Computer-Vision-on-AWS`

A Jupyter notebook is available for running the example code from this chapter. You can access the most recent code from this book's GitHub repository, `https://github.com/PacktPublishing/Computer-Vision-on-AWS`. Clone that repository to your local machine using the following command:

```
$ git clone https://github.com/PacktPublishing/Computer-Vision-
on-AWS
$ cd Computer-Vision-on-AWS/03_RekognitionCustomLabels
```

Additionally, you will need an AWS account and Jupyter notebook. *Chapter 1* contains detailed instructions for configuring the developer environment.

Introducing Amazon Lookout for Vision

As we discussed in *Chapter 3*, building a custom ML model from scratch is an overwhelming undertaking that requires tremendous time, not to mention that it requires thousands of images along with expert ML practitioners.

With Amazon Lookout for Vision, you don't need to invest your time and resources in building ML or deep learning pipelines. Instead, you just need to provide normal and anomalous images and the service does the rest. If your images are already labeled, you can directly import them; otherwise, you can use the service's built-in labeling interface or utilize **SageMaker GroundTruth** for labeling. Once you have labeled data, you just need to initiate single-click training. Once the model training completes, you will receive results showing the model's performance.

You can get started with Lookout for Vision with few images (as few as 30 images) but you may want to provide a larger dataset for complex use cases requiring high accuracy. When you're satisfied with the model's accuracy, you can start hosting the trained model with a single click. Amazon Lookout for Vision also allows the deployment of trained models on edge devices managed by **AWS IoT Greengrass Version 2**. We'll dive deeper into this use case in *Chapter 8*.

The benefits of Amazon Lookout for Vision

Amazon Lookout for Vision provides similar benefits to Amazon Rekognition Custom Labels, but is specifically focused on CV-based inspection in industrial processes:

- **Ease of adoption**: Amazon Lookout for Vision provides easy access to CV-based ML. You do not need ML expertise to build your model. As a user, you just need to bring sample baseline images and Lookout for Vision will automatically load and inspect the data, choose the most suitable algorithm, train a custom ML model, and provide model performance metrics.

- **Optimize production quality and operational cost**: As Lookout for Vision can identify anomalies, you can take quick action to avoid defects and increase production quality. You can take it a step further by analyzing trends that flag processes causing the anomalies or updating the maintenance routine.

Now that we have gone through the basics of Amazon Lookout for Vision, we will learn how to use it to create models for our use cases.

Creating a model using Amazon Lookout for Vision

In this section, we will learn how to choose the type of model you need for your use case, how to create training and test datasets, start model training, improve your model, and analyze images using the trained model.

Choosing the model type based on your business goals

Similar to Rekognition Custom Labels, Lookout for Vision provides a couple of choices in choosing the model type: image classification and image segmentation:

- **The image classification model**: This type of model predicts whether the image contains an anomaly or not. It will not provide other information, such as the location of the anomaly or the type of anomaly. An example of this is if you want to know whether a capsule shell is mashed or cracked on an assembly line. An image classification model would be a good choice for this use case. As capsule shells are relatively cheap, you may not want to investigate the location of mashing or cracks.

- **The image segmentation model**: This type of model predicts the classification of whether the image contains an anomaly or not, and segmentation information such as the location, the type of anomaly, and the percentage area of the image that the anomaly covers. For example, if you need to determine the type and level of damage to a vehicle, you can leverage the image segmentation model. It can help you determine whether the vehicle has any damage or not (the presence of an anomaly). If there is any damage (an anomaly), it will capture the type of damage (a scratch, a dent, and so on), the location of the damage, and the area of the damage (to help you determine whether it is a minor or major repair).

Creating a model

Once you have decided on the type of model you need, you can start creating a model with Amazon Lookout for Vision. You will need to create a project, provide datasets (images), and initiate the training process for the model.

Creating a project

A project is a group of resources needed to create and manage datasets and models. A project contains and manages the following:

- **Datasets**: A project has training and test datasets (images) and image labels that are used to train a model.

- **Models**: A project contains ML models you train to find scenes, objects, and their locations. You can have more than one version of a model in a project.

Each project must contain resources for individual use cases, such as classifying whether a **printed circuit board** (**PCB**) is anomalous or normal or identifying types and locations of anomalies for a machine part.

Creating datasets

To train a model with Lookout for Vision, you need to provide images of normal and anomalous objects for your use case. These images are supplied to the service as a dataset. Within your project, you create a training dataset and a test dataset that the service uses to train and test your model. You can create training and test datasets using the Lookout for Vision console or AWS **software development kit** (**SDK**).

Creating training and test datasets using the AWS Management Console

You can start a project with a single dataset or separate training and test datasets. AWS recommends using a single dataset for a project unless you want granular control over training and testing. Once you have created a single dataset, the service will split your dataset during training to create a training dataset (80%) and a test dataset (20%) for your project.

To create the datasets for a project, you can import the images in one of the following ways:

- Import images from your local computer.

- Import images from an S3 bucket. Amazon Lookout for Vision can label the images using the folder names that contain the images, such as normal for good images and anomaly for anomalous images.

- Import an Amazon SageMaker Ground Truth manifest file. We will cover how to label data with SageMaker Ground Truth in *Chapter 9*.

Depending on where you import your images from, your images might not be labeled. You can use the Lookout for Vision console to add, change, and assign labels. You do need to label the dataset images according to the type of model you plan to use (classification versus segmentation). The labels determine the model type that Lookout for Vision will create.

There are a few other things to keep in mind while creating datasets for a project:

- All images in a dataset must have the same dimension (otherwise, model training will fail).

- Each of these images must contain only one type of anomaly. Providing images with multiple anomalies in a single image will reduce the model performance and increase the number of incorrect predictions.

- Try to provide images with consistent conditions, such as the position of the camera (the same angle, zoom, quality, and so on), lighting, and pose of the object.

- If you are providing separate datasets for training and testing, make sure to avoid duplicates.

Training the model

Once you have created training and test datasets, you can initiate the model training. Keep in mind that a new version of a model is created every time you initiate training. If you provided a single dataset, the service will split it into training and test datasets before it starts training. After the training is complete, you can use the results to evaluate and improve your model. Training can take some time to complete. You are only charged for successful model training. If model training fails, the service will provide debugging information for you to use.

Improving the model

Once the training completes, you can evaluate the model using the metrics provided by Lookout for Vision. For image classification models, Lookout for Vision provides metrics such as precision, recall, and an F1-score. For image segmentation models, you will receive image classification metrics (precision, recall, and an F1 score) and additional performance metrics (an F1 score and an average Intersection over Union score) for each individual anomaly label. We covered precision, recall, and F1 metrics in detail in *Chapter 3*.

> **Important note**
> The average **Intersection over Union** (**IoU**) is the average percentage overlap between the anomaly mask (the area of an anomaly – similar to the bounding box in Custom Labels) in test images and the anomaly mask predicted by the model; a high value indicates the model is performing well.

Amazon Lookout for Vision also supports a specific model improvement feature – a trial detection task. A trial detection task will identify anomalies in new images (not the training/test dataset) you provide; you can think of it as a trial run. Once the trial detection task is run, you can verify the result and correct predictions if necessary. After verifying the images, you can add them to the training dataset and then retrain your model. You should run trial detection tasks periodically to continuously improve your model quality and performance.

Starting your model

Once you're happy with the performance of the model, you need to start the model before you can use it. You can start the model by using the `StartModel` API. You need to provide the capacity (known as inference units) to provision the necessary compute resources. You will be charged based on the allocated capacity for the amount of time that your model runs.

Analyzing an image

Once the model is hosted, you can start analyzing your images using the `DetectAnomalies` API. As part of the API call, you can supply an image along with the project name and model version that you want to use.

Stopping your model

Since you are charged for the time that your model is running, you should stop the model if it is not in use. You can stop the model by using the `StopModel` API.

Now that we have gone through the steps of building a model with Amazon Lookout for Vision, let's learn how to train and evaluate a model to identify damaged pills.

Building a model to identify damaged pills

In this section, we will learn how to train a model to identify damaged pills using Amazon Lookout for Vision. We'll collect training and test datasets for pills, label the images, and train a model. We will use an image classification model for this use case to identify whether the pills are damaged or not. We will not classify the anomaly into categories such as chipped or cracked, or highlight the areas of an anomaly in this example.

Step 1 – collecting your images

1. Upload the sample images from the book's GitHub repository. You can complete this step using the following command:

```
$ aws s3 sync 07_LookoutForVision/images s3://cv-on-aws-book-
xxxx/chapter_07/images --region us-east-2
```

> **Important note**
> To collect the sample images, you can use the same S3 bucket you created in *Chapter 2*.

2. Now, navigate to Amazon Lookout for Vision on the AWS Management Console (`https://us-east-2.console.aws.amazon.com/lookoutvision?region=us-east-2#/`).

3. Select on **Get started**:

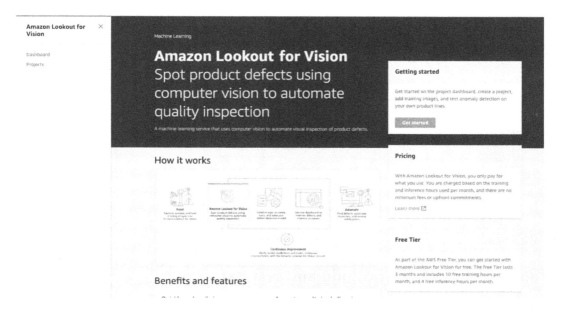

Figure 7.1: The Amazon Lookout for Vision console

> **Important note**
> If you're a first-time user of the service, it will ask permission to create an S3 bucket to store your project files. Select **Create S3 bucket**.

Step 2 – creating a project

Next, selct on **Create project**.

Figure 7.2: Creating a project using the Lookout for Vision console

Give it a name, such as `pills-damage-detection`.

Step 3 – creating the training and test datasets

Next, as we saw in the *Creating a model* section, we need to create a dataset. Select on **Create dataset**:

Figure 7.3: Creating a dataset on the Lookout for Vision console

On the **Create dataset** page, leave the default configuration option of creating a single dataset (the service will split the dataset into training and test datasets).

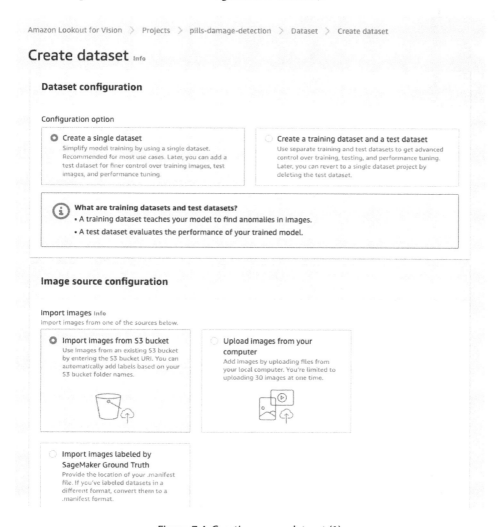

Figure 7.4: Creating a new dataset (1)

Under **Image source configuration**, select **Import images from S3 bucket** and provide the S3 URI for your bucket where you copied the images in *step 1*. The S3 URI will look similar to s3://cv-on-aws-book-xxxx/chapter_07/images/.

Important note

Keep in mind to add the trailing / in the S3 URI.

As our images are organized appropriately into normal and anomaly folders, we will check the **Automatic labeling** checkbox for this project. This option will automatically label the images after they're imported. Select on **Create dataset**.

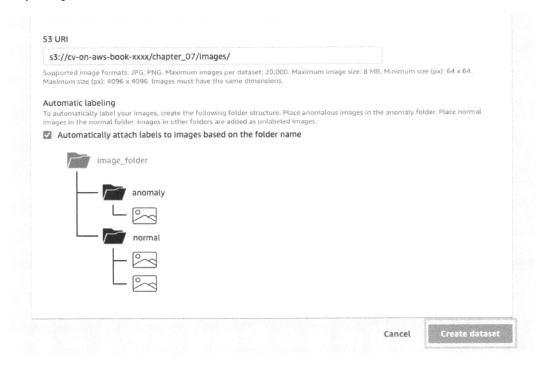

Figure 7.5: Creating a new dataset (2)

Step 4 – verifying the dataset

Once the images have been imported into the project, we can verify whether the images are correctly labeled. As we used the auto-labeling feature, we don't need to label the images, but if you import unlabeled images, you will need to label the images before proceeding further:

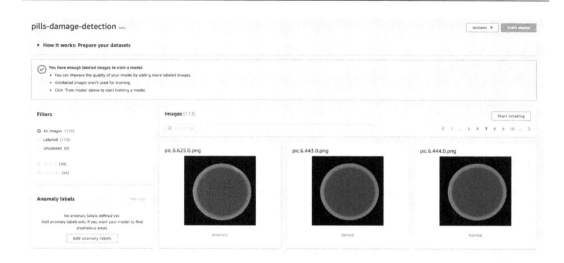

Figure 7.6: Verifying the dataset

Optionally, if you want to make any changes to the labeled images (or label unlabeled images), select on **Start labeling**. Once you're in edit mode, select the images you want to label.

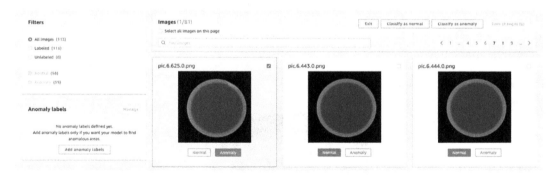

Figure 7.7: Label images

Optionally, if you want your model to find anomalous areas, you can select on **Add anomaly labels**.

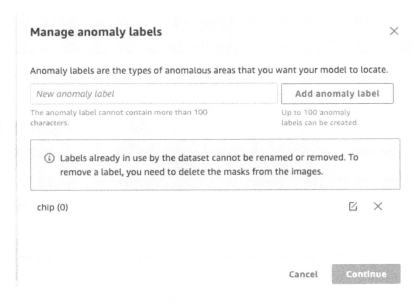

Figure 7.8: Creating new anomaly labels

If you need to find anomalous areas, you can use the annotation tool. You can select on the images to access the annotation tool:

Figure 7.9: Assigning an anomaly label and drawing an anomaly mask

> **Important note**
> You can use the additional capabilities of the labeling panel below the image to adjust or correct your bounding boxes or labels.

For our project, we will not add any new labels or draw any anomaly masks. We just want to identify whether the pill has any damage or not. If you did make any changes that you would like to revert, do not save those changes; instead, select on **Exit**. If you created new anomaly labels, delete them.

Step 5 – training your model

Once we have completed verifying all the images, the next step is to train a model. Select on **Train model**:

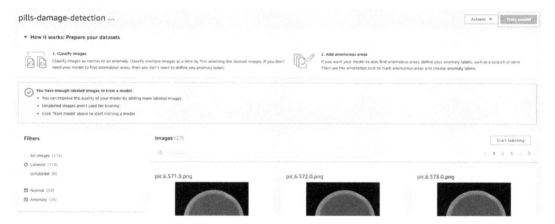

Figure 7.10: Train model (1)

As we discussed, the service takes care of the heavy lifting from a model training perspective so all you need to do is select on **Train model**:

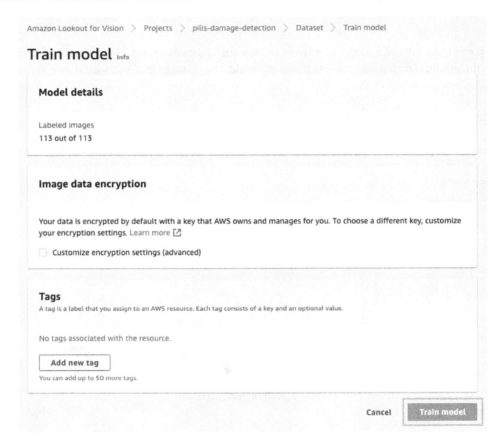

Figure 7.11: Train model (2)

> **Important note**
> You can add tags if you'd like to track your models. By default, your data is encrypted with a default key that AWS owns and manages for you. However, you can also encrypt your images with your own encryption key through Amazon Key Management Service.

The console will ask you to confirm whether you want to proceed with training; select on **Train model** to proceed. You will see the **Status** value set to **Training in progress**.

The model training will take typically from 30 minutes to 24 hours to complete. You will be charged for the amount of time it takes to successfully train your model.

Validating it works

Once you see the status model changed to **Training complete**, select on the model name (**Model 1**). You will see performance metrics and an overview of the test results. As you can see in the following screenshot, our trained model has a really good F1 score, precision, and recall:

Figure 7.12: Evaluating the model performance

Step 1 – trial detection

As we previously discussed, Lookout for Vision allows you to create trial detection tasks and verify results. Afterward, if needed, you can retrain your model to improve the results.

Let's upload images for trial detection from the book's GitHub repository. You can complete this step using the following command:

```
$ aws s3 sync 07_LookoutForVision/trial_detection s3://cv-on-
aws-book-xxxx/chapter_07/trial_detection --region us-east-2
```

Select on **Trial detections** on the pane on the left, and then Select **Run trial detection**:

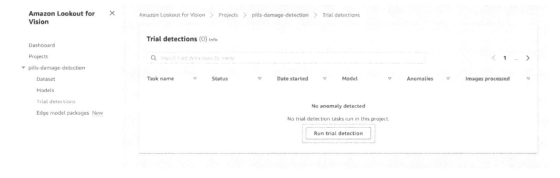

Figure 7.13: Run trial detection (1)

Next, provide the task name, and choose the model and location of images for trial detection. Select **Detect anomalies**:

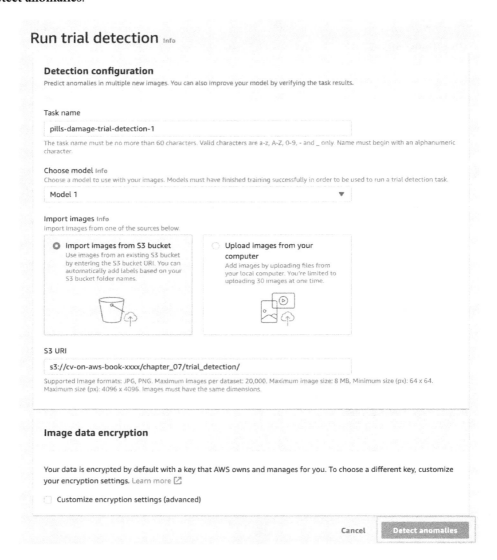

Figure 7.14: Run trial detection (2)

When you are asked to confirm that you want to run a trial detection task, select on **Run trial detection**. Once the task is complete, the status will change to **Detection complete**. Select on the task name and it will show the predictions and the option to verify the results:

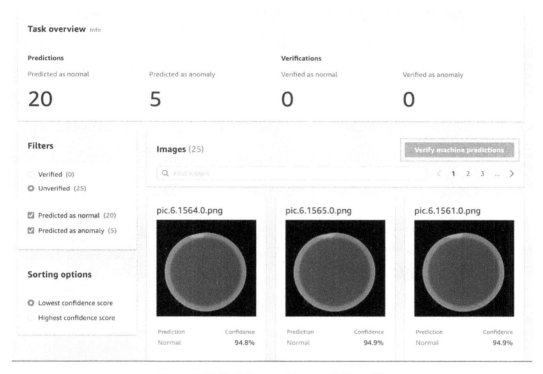

Figure 7.15: Verifying machine predictions (1)

Next, we need to verify the machine predictions to confirm that the image classifications are correct. Select on **Verify machine predictions**:

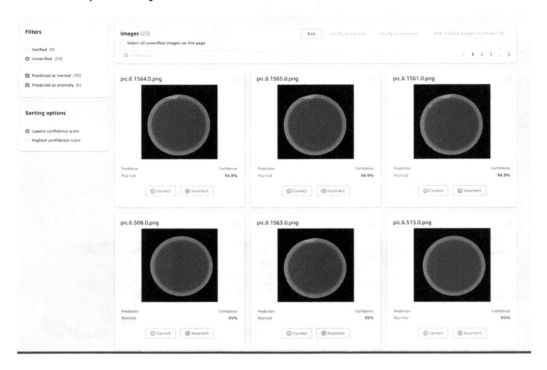

Figure 7.16: Verifying machine predictions (2)

Mark each image as correct or incorrect depending on whether the machine predictions are accurate:

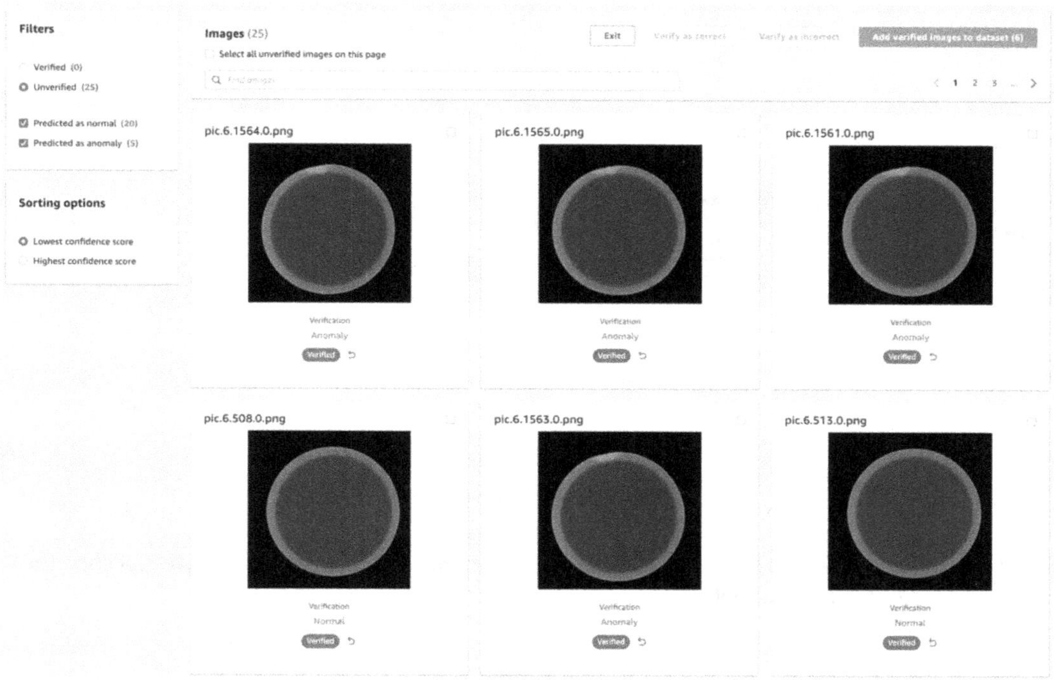

Figure 7.17: Verifying machine predictions (3)

After all the images are verified, select on **Add verified images to dataset**. Make sure your S3 bucket has versioning enabled for the service to update the manifest file; otherwise, you will receive an error.

Trial detection will add images to your project dataset and you can retrain your model to improve performance if needed. For our project, we'll skip retraining the model and will go ahead with starting the model.

Step 2 – starting your model

The next step is to start the model so we can start using it to detect damage in images of pills. To do this, select on the **Use model** tab and then **Integrate API to the cloud**:

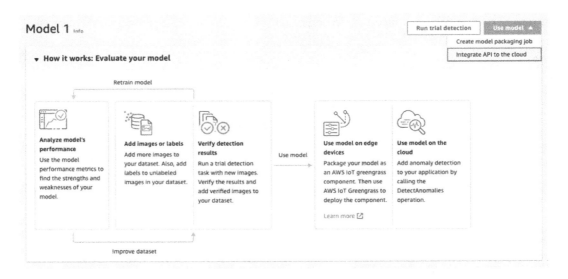

Figure 7.18: Using the model

You will be directed to a page containing AWS CLI commands for various actions. We can copy the commands under the **Start model** title:

Figure 7.19: Starting the model

As you can see, by default, it selects 1 inference unit. You may need to provide a --region parameter if your AWS CLI is not configured with the default region, us-east-2:

```
aws lookoutvision start-model --project-name pills-damage-
detection --model-version 1 --min-inference-units 1 --region
us-east-2
```

Once you execute the StartModel command, you will see **Status** set to **Starting hosting**.

> **Important note**
> A higher number of inference units will increase the throughput of your model, allowing you to process a greater number of images per second.

Step 3 – analyzing an image with your model

Once the model is hosted, you can use it via the AWS CLI or the open source Python SDK (developed specifically for Lookout for Vision).

We will use the AWS CLI command provided through the Lookout for Vision console. We provide the image type and location of the image (`pic.7.615.0.png`):

```
aws lookoutvision detect-anomalies \
  --project-name pills-damage-detection \
  --model-version 1 \
  --content-type image/png \
  --body /path/to/pic.7.615.0.png\
  --region us-east-2
```

The response to the preceding command will look like the following:

```
{
    "DetectAnomalyResult": {
        "Source": {
            "Type": "direct"
        },
        "IsAnomalous": false,
        "Confidence": 0.7335836887359619
    }
}
```

The response conveyed that the trained model did not detect any damage to the pill in the image and the result had a confidence score of 73.35%.

The Amazon Lookout for Vision console also provides a dashboard that shows metrics for your projects, such as the total number of images processed, the total number of anomalies detected, and the ratio and anomaly count over a period of time.

Figure 7.20: The Amazon Lookout for Vision dashboard

Step 4 – stopping your model

You incur charges while the model is running. You should stop your model when it's not being used:

```
aws lookoutvision stop-model --project-name pills-damage-
detection --model-version 1 --region us-east-2
```

In this section, we learned about how we can run trial detection on a trained model to validate its performance and retrain the model if needed. Then, we used the trained model to analyze an image to detect damaged pills.

Summary

In this chapter, we covered what Amazon Lookout for Vision is and how you can use it for automated quality inspection to detect anomalies, such as damage, irregularities, defects, or missing component in images. We discussed the benefits it provides and the process to create a model using the service. Finally, we trained and deployed a model to detect damaged pills.

In the next chapter, we will learn about how we can take this further and use it to detect manufacturing defects in the manufacturing facility by deploying CV models at the edge.

Detecting Manufacturing Defects Using CV at the Edge

In the previous chapter, we introduced Amazon Lookout for Vision. We discussed how you can use Lookout for Vision to detect defects at scale using CV and walked through an example that trained a model to identify damaged pills.

In this chapter, we will dive deeper into Lookout for Vision by detecting manufacturing defects in images of parts while running models on edge devices. What are edge devices? With edge computing, you can run applications away from the cloud and locally on computer devices (edge devices). Some examples of edge devices are security cameras, smartphones, and sensors. There are a variety of use cases for edge computing, including medical imaging, predictive maintenance, and defect detection. Detecting defects early in the manufacturing process can help with improved quality control and enhanced safety measures. One benefit of deploying ML models on edge devices is that monitoring can be enabled to detect defects in real time. This helps prevent machine failure and provides cost savings by reducing the need to remanufacture parts due to flaws. In the next section, we will discuss the benefits and challenges of running ML models on the edge.

In this chapter, we will cover the following topics:

- Understanding ML at the edge
- Deploying a model at the edge using Lookout for Vision and AWS IoT Greengrass

Technical requirements

You will require the following:

- Access to an active AWS account with permissions to access Amazon Lookout for Vision and AWS IoT Greengrass Version 2 (V2)
- The code examples for this chapter, which can be found on GitHub at `https://github.com/PacktPublishing/Computer-Vision-on-AWS/tree/main/08_EdgeDeployment`

Understanding ML at the edge

Throughout this book, we have discussed many CV use cases and provided steps to deploy models on AWS for inference. What if these models are unable to be deployed to the cloud? To perform model training, we often deal with vast amounts of data and large-sized images. In some instances, you may encounter legal restrictions or security regulations that require this data to be processed on-premises. You may also be processing video from a camera on a manufacturing floor and discover that there are bandwidth constraints that limit how much video content is sent to the cloud. In healthcare, medical images that are generated by CT scans and MRI machines provide crucial information for clinicians and providers. These images need to be processed in near real time for quick delivery of the right treatment plan to a patient. Uploading the images to the cloud would incur high latency and delay the ML model insights used to help make clinical decisions. Running ML models locally on edge devices ensures lower latency for real-time predictions.

It is challenging to create a robust and secure infrastructure for ML at the edge. Installing big frameworks such as TensorFlow or PyTorch on an edge device may not be feasible due to restrictions on the device and storage capacity limitations. A model's performance needs to be monitored and retrained if needed. The model also needs to be secured on the device. There are scalability challenges for ML at the edge, such as operating on and deploying models to multiple edge devices. AWS-managed services address these challenges and help meet your edge computing requirements.

Using Amazon Lookout for Vision eliminates the need to install a large ML framework and allows you to deploy your model to an edge device. This helps with lower bandwidth costs and performing real-time image analysis locally. The models can be deployed on several supported devices and architectures, including NVIDIA Jetson Xavier and x86 platforms running Linux. You can also use **AWS IoT Greengrass Version 2 (V2)**, which is an **Internet-of-Things (IoT)** edge service for building, running, managing, and deploying IoT applications on many supported edge devices. It supports running its core software on x86 and ARM architectures. You can deploy a Lookout for Vision model on an AWS IoT Greengrass V2 core device. The requirements can be reviewed in the documentation (`https://docs.aws.amazon.com/lookout-for-vision/latest/developer-guide/models-devices-setup-requirements.html`):

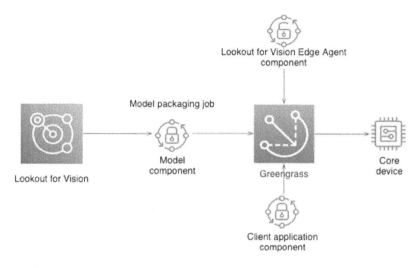

Figure 8.1: Amazon Lookout for Vision and AWS IoT Greengrass edge deployment overview

In the next section, we will go through a hands-on example to detect manufacturing defects. We will use images of a circuit board to train a model that identifies whether anomalies exist in the images. In this example, we will upload our dataset to S3 to train our model. If you require managed hardware for your edge computing needs, there are options such as AWS Outposts, AWS Storage Gateway, and the AWS Snow Family. Visit https://aws.amazon.com/what-is/edge-computing/ to view more information about the available edge computing options. Another option, if you are unable to send your images to the AWS cloud, is to use an AWS Panorama Appliance (https://aws. amazon.com/panorama/), where you can run CV applications at the edge.

Deploying a model at the edge using Lookout for Vision and AWS IoT Greengrass

In this section, we will learn how to deploy a model that was trained using Lookout for Vision to an edge device using AWS IoT Greengrass V2. For this hands-on example, we will create an EC2 instance to simulate an edge device.

Step 1 – Launch an Amazon EC2 instance

The first step is to launch an EC2 instance where we will install AWS IoT Greengrass V2. We will use an Ubuntu 20.04 c5.2xlarge instance. Now, navigate to EC2 via the AWS Management Console (https://us-east-1.console.aws.amazon.com/ec2/home?region=us-east-1#Home). Select on **Launch instance**:

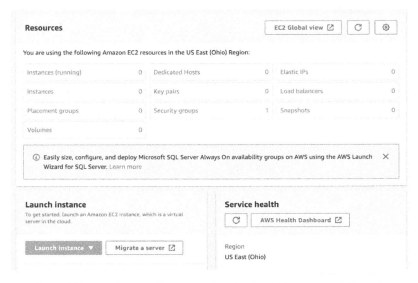

Figure 8.2: Launching an instance using the AWS Management Console

Enter a **Name** value and under the **Quick Start** section, select the **Ubuntu Server 20.04** Amazon Machine Image. Confirm that the architecture is x86. Under **Instance type**, select **c5.2xlarge**. Keep the default values for the rest of the configuration details and select **Launch instance**. You are not required to create a key pair, since we will be using EC2 Instance Connect in a later step to connect to our EC2 instance:

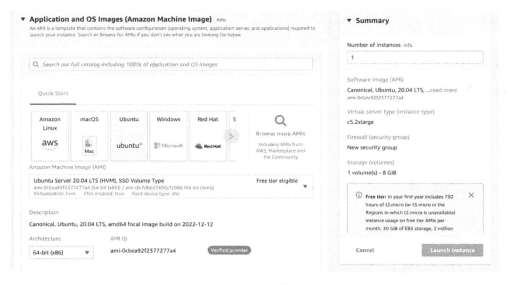

Figure 8.3: EC2 instance configuration details

Step 2 – Create an IAM role and attach it to an EC2 instance

Before connecting to the EC2 instance and installing AWS IoT Greengrass V2, we need to create a new IAM role with the required permissions. Navigate to IAM (`https://us-east-1.console.aws.amazon.com/iamv2/home?region=us-east-1#/roles`) and select **Create role**:

Figure 8.4: The IAM console to create a new role

On the **Select trusted entity** page, select **AWS service** under **Trusted entity type** and **EC2** under **Use case**, then select **Next**:

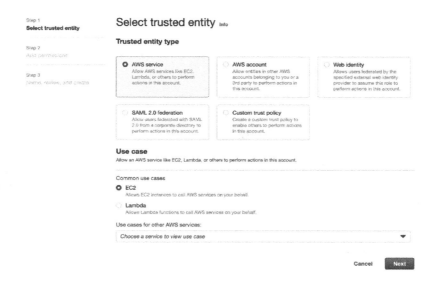

Figure 8.5: Using the Select trusted entity page to allow an EC2 service to assume a new IAM role

On the **Add permissions** page, attach the following policies and select **Next**:

- **AmazonEC2FullAccess**
- **AmazonSSMFullAccess**
- **AWSGreengrassFullAccess**
- **AWSIoTFullAccess**

- **AmazonLookoutVisionFullAccess**

- **AmazonS3FullAccess**

- **CloudWatchAgentAdminPolicy**

- **IAMFullAccess**

Create a role name and add a description for the role, then select **Create role**:

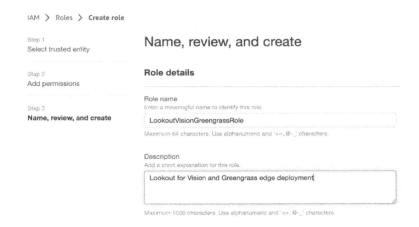

Figure 8.6: Creating a new IAM role

Now that the new IAM role has been created, navigate back to the EC2 console to modify the IAM role. Check the checkbox next to the **Edge** instance and, from the **Actions** dropdown, select **Security > Modify IAM role**:

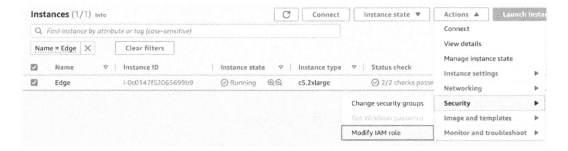

Figure 8.7: Modifying the IAM role for the EC2 instance

On the **Modify IAM role** page, update the IAM role to the new role we just created and select **Update IAM role**.

Step 3 – Install AWS IoT Greengrass V2

Once the instance is running, check the checkbox next to the instance and select **Connect**:

Figure 8.8: Connecting to the EC2 instance

Use **EC2 Instance Connect** to connect to the instance to install AWS IoT Greengrass V2 and the required dependencies. Select on **Connect**. A new tab will open where you will access the Terminal:

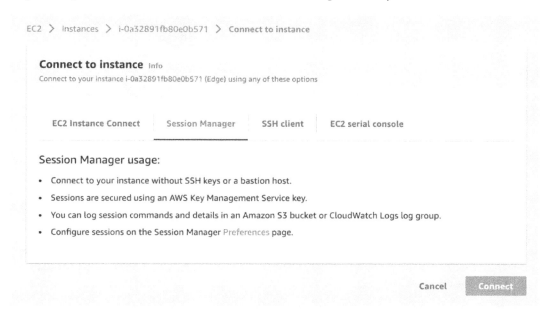

Figure 8.9: Using EC2 Instance Connect to connect to the instance

Clone the GitHub repository:

```
$ git clone https://github.com/PacktPublishing/Computer-Vision-
on-AWS
$ cd Computer-Vision-on-AWS/08_EdgeDeployment/edge
```

There are a few scripts that reside under the edge directory. Let's review their purpose:

- To call the Lookout for Vision Edge Agent, you need to set up a gRPC client. The `.proto` gRPC stub file is provided by AWS (https://docs.aws.amazon.com/lookout-for-vision/latest/developer-guide/client-application-overview.html), which creates the `edge_agent_pb2_grpc.py` and `edge_agent_pb2.py` Python client interfaces
- The install scripts install the required packages and the AWS IoT Greengrass Core software (https://docs.aws.amazon.com/greengrass/v2/developerguide/quick-installation.html)
- The remaining scripts are used to interact with the Lookout for Vision model and to detect anomalies in the circuit board images

Update the permissions for the install scripts to make them executable and install the ZIP package and the boto3 AWS SDK:

```
$ chmod a+x install*
$ sudo apt-get install zip
$ pip3 install boto3
```

Run the following commands to install AWS IoT Greengrass V2 and the necessary Python dependencies:

```
$ ./install_greengrass.sh
$ ./install-ec2-ubuntu-deps.sh
$ pip3 install -r requirements.txt
```

Confirm Greengrass is running by navigating to **AWS IoT Core** > **Manage** > **Greengrass devices** > **Core devices** (https://us-east-1.console.aws.amazon.com/iot/home?region=us-east-1#/greengrass/v2/cores):

Figure 8.10: AWS IoT Greengrass core devices

Step 4 – Upload training and test datasets to S3

Upload the sample images from this book's GitHub repository. You can complete this step using the following command:

```
$ aws s3 sync 08_EdgeDeployment/images s3://cv-on-aws-book-
xxxx/08_EdgeDeployment/images --region us-east-1
```

> **Important note**
> To collect the sample images, you can use the same S3 bucket you created in *Chapter 2*.

Step 5 – Create a project

Navigate to Amazon Lookout for Vision on the AWS Management Console (https://us-east-1. console.aws.amazon.com/lookoutvision?region=us-east-1#/). Select on **Create project**:

Figure 8.11: Creating a project using Lookout for Vision

Provide a name, such as `EdgeDemo`.

Step 6 – Create training and test datasets

We need to create training and test datasets to train the model and perform inference, so select on **Create dataset**:

Figure 8.12: Creating datasets in Lookout for Vision

Select **Create a training dataset and a test dataset** and **Import images from S3 bucket**. Provide the S3 URI for the bucket where you copied the images in *step 3*. The S3 URI will look similar to `s3://cv-on-aws-book-xxxx/08_EdgeDeployment/images/train/` for the training dataset and `s3://cv-on-aws-book-xxxx/08_EdgeDeployment/images/test/` for the test dataset:

> **Important note**
> Keep in mind that you must add the trailing / in the S3 URI.

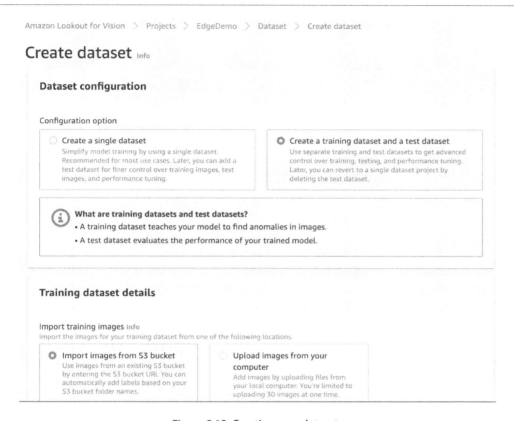

Figure 8.13: Creating new datasets

Since our images have been organized appropriately into normal and anomalous folders within the test and train folders, we will check the **Automatic labeling** checkbox for this project. This option will auto-label the images after they're imported. Select on **Create dataset**.

Step 7 – Train the model

Once the images have been imported into the project, we can verify whether the images have been labeled correctly. Since we used the auto-labeling feature, we don't need to label the images. Next, select on **Train model**:

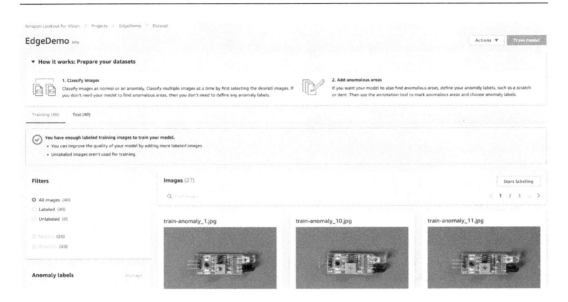

Figure 8.14: Training the model

The model training will take approximately 45-60 minutes to complete. Once the training process is complete, review the **Model performance metrics** and **Test results overview** areas:

Figure 8.15: Model performance metrics and Test results overview areas

Step 8 – Package the model

To deploy the model on the EC2 instance, we need to package the model as an AWS IoT Greengrass component. Navigate to the model on the Lookout for Vision console. Then, select **Use model** and select **Create model packaging job**:

Figure 8.16: Create model packaging job

Provide a job name and select the model that you trained in the previous step. In the **Target hardware settings** section, select the **Target platform** radio button and enter **X86_64** for **Architecture**, CPU for **Accelerator**, and **{"mcpu": "core-avx2"}** for **Compiler options**:

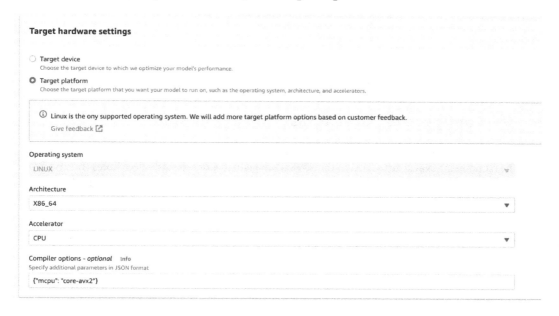

Figure 8.17: Selecting the target hardware settings

On the **AWS IoT Greengrass component settings** page, enter a **Component name** value and select an S3 bucket where you will store the component artifacts. Then, select on **Create model packaging job**:

AWS IoT Greengrass component settings

Component name   Info
Enter a name for your model's AWS IoT Greengrass component. The name identifies the component within the AWS IoT Greengrass console.

> Enter component name

The component name must be no more than 128 characters. Valid characters are a-z, A-Z 0-9, -, _ and .

Component description - *optional*
Enter a description for your component.

> Enter component description

The description must be no more than 256 characters

Component version   Info
Enter a component version. If you don't specify a version, Lookout for Vision derives a version from the version of your model.

> 1.0.0

AWS IoT Greengrass uses semantic versioning scheme for components. Learn more 🗗

Component location   Info
Enter the S3 location in which Amazon Lookout for Vision stores the component artifacts.

> Q s3://bucket/prefix/   View 🗗   Browse S3

Figure 8.18: AWS IoT Greengrass component settings

Step 9 – Configure IoT Greengrass IAM permissions

The IoT Greengrass IAM role requires permissions to access the model artifact in S3 before we can deploy the component. Navigate to IAM in the AWS Management Console and select **Roles** (`https://us-east-1.console.aws.amazon.com/iamv2/home?region=us-east-1#/roles`). Search for the IAM role that was created earlier – for example, `MyGreengrassV2Role`:

IAM > Roles

Roles (Selected 1/122) Info
An IAM role is an identity you can create that has specific permissions with credentials that are valid for short durations. Roles can be assumed by entities that you trust.

Q greengrass	✕	1 match
☑ **Role name**	▽	**Trusted entities**
☑ MyGreengrassV2Role		AWS Service: credentials.iot

Figure 8.19: AWS IoT Greengrass IAM role

Select on the role. Under **Permissions**, select **Add Permissions** > **Attach policies**, then attach the `AmazonS3FullAccess` managed policy:

Figure 8.20: S3 access added to the AWS IoT Greengrass IAM role

Step 10 – Deploy the model

Navigate back to the Lookout for Vision console and select the model packaging job. Select on **Continue deployment in Greengrass**. This will open a new tab in the AWS IoT console:

Figure 8.21: Deploying the model package job

Select **Deploy** and select **Create new deployment**:

Figure 8.22: Adding a new deployment

Enter a name in the **Deployment target** section. Then, select **Core device** and enter the core device name we provisioned earlier:

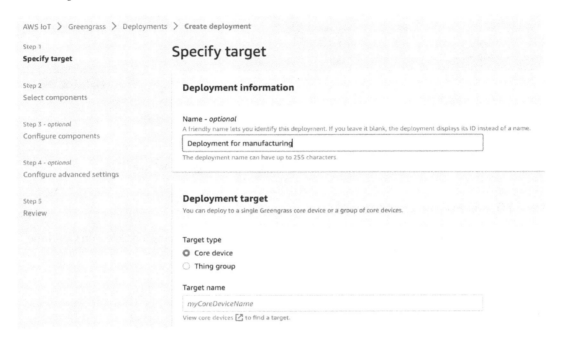

Figure 8.23: Model deployment target

Select **Next**. Select the following public components to better interact with the model on the device:

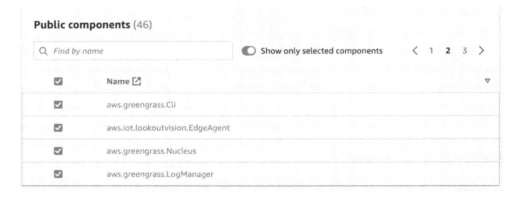

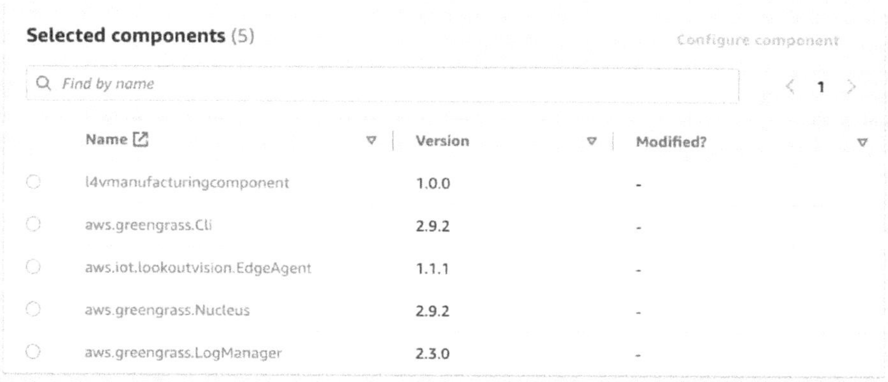

Selected components (5)   Configure component

Q Find by name			‹ 1 ›
Name ⌑ ▽	**Version** ▽	**Modified?** ▽	
○ l4vmanufacturingcomponent	1.0.0	-	
○ aws.greengrass.Cli	2.9.2	-	
○ aws.iot.lookoutvision.EdgeAgent	1.1.1	-	
○ aws.greengrass.Nucleus	2.9.2	-	
○ aws.greengrass.LogManager	2.3.0	-	

Figure 8.24: Components to deploy to the device

Select on **Next**, review the configuration, and select on **Deploy**.

Step 11 – Run inference on the model

Once the model deployment is successful, navigate back to the EC2 instance Terminal using Session Manager, as shown earlier in *step 2*. Run the following command to confirm all the components are running:

```
$ sudo /greengrass/v2/bin/greengrass-cli component list
```

Before we perform inference, we must load the model:

> **Important note**
> Confirm you're running the commands under the correct directory.

```
$ cd edge
$ python3 warmup-model.py <model component name>
```

Now, we're ready to perform inference on the model using our test images. Navigate to the test.py script and edit the ENDPOINT, CLIENT_ID, and MODEL_COMPONENT parameters. CLIENT_ID can be found by navigating to the AWS IoT Core console, choosing **Manage** > **All Devices** > **Things**, and copying the Thing's name – for example, l4vEdgeDemoThing:

Figure 8.25: Thing name for the CLIENT_ID parameter

Select on the Thing's name, select the **Interact** tab, and select on **View Settings**. Under the **Device data endpoint** section, copy the endpoint and replace it in the ENDPOINT parameter:

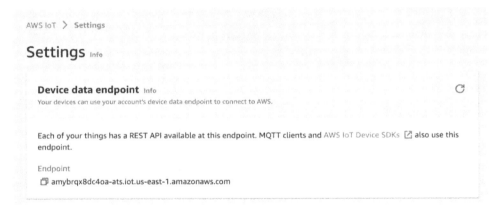

Figure 8.26: Endpoint name for the ENDPOINT parameter

Run the following command to generate inference using one of the anomalous images:

```
$ python3 test.py <path to test image>
```

The output should look similar to the following, where the result is also published as an MQTT message:

```
run inference on model component "l4vmanufacturingcomponent"
with image ../images/test/anomaly/test-anomaly_2.jpg
```

```
input image shape: (2667, 4000, 3)
the image has an anomaly - confidence 0.7299103140830994
Connecting to xxxx.us-east-2.amazonaws.com with client ID
"l4vEdgeDemoThing"...
Connected!
Begin Publish
Published: '{"message": "is_anomalous: true\nconfidence:
0.7299103140830994\n [1]", "is_anomalous": true, "confidence":
0.7299103140830994}' to the topic: l4vmanufacturing/testclient
Published End
```

Step 12 – Clean up resources

To prevent yourself from incurring additional charges, terminate the EC2 instance, go to the Lookout for Vision console and delete the project, and, in the IoT Greengrass console, select your deployment, and select on **Cancel**.

Summary

In this chapter, we explained the purpose of edge computing. We also covered the benefits and challenges of ML at the edge. We walked through a code sample to identify defects in images using Amazon Lookout for Vision and deployed the trained model to an edge device using AWS IoT Greengrass V2. In the next chapter, we will introduce Amazon SageMaker Ground Truth and discuss how it can be used to quickly label data using various workforce options.

Part 4: Building CV Solutions with Amazon SageMaker

This fourth part consists of two cumulative chapters that will cover how to use Amazon SageMaker for CV use cases.

By the end of this part, you will understand how to use Amazon SageMaker Ground Truth to accelerate your labeling jobs and integrate a human labeling job into your offline data labeling workflows. You will also walk through a code example that uses Amazon SageMaker to train a model using a built-in image classifier.

This part comprises the following chapters:

- *Chapter 9, Labeling Data with Amazon SageMaker Ground Truth*
- *Chapter 10, Using Amazon SageMaker for CV*

9
Labeling Data with Amazon SageMaker Ground Truth

Labeled data is key to developing an accurate and effective model using a supervised machine learning algorithm. Typically, machine learning practitioners spend 70% of their time labeling and managing data. It slows down innovation and increases cost. We saw in *Chapter 3*, and *Chapter 7*, how we needed high-quality labeled data to develop custom ML models. Although those services allowed a labeling interface in the built-in console, if you have a large number of images in your dataset, it can quickly become a monumental and cumbersome task to label them. You would either need to outsource the labeling responsibility or would need a solution to split the labeling workload across multiple labelers. Building such a solution is undifferentiated heavy lifting for ML practioners who would just want to focus on developing accurate models.

The solution to this labeling challenge is Amazon SageMaker **Ground Truth** (**GT**). As the name of the service suggests, it helps build ground truth (accurately labeled data) for your machine learning models. You can leverage SageMaker GT to create a managed data labeling pipeline where you can ingest raw, unlabeled data, send it to human labelers, and store the output (labeled data) in a data store. Using Ground Truth, you can use external workers or you can use your internal workforce to enable you to create a labeled dataset.

In this chapter, we will cover the following topics:

- How to build a data labeling job
- How to label a set of images with a private workforce
- Importing the labeled data into a Rekognition Custom Labels project

Technical requirements

You will require the following:

- Access to an active AWS account with permissions to access Amazon SageMaker and Amazon Lookout for Vision

- PyCharm or any Python IDE

- All the code examples for this chapter can be found on GitHub at `https://github.com/PacktPublishing/Computer-Vision-on-AWS`

A Jupyter notebook is available for running the example code from this chapter. You can access the most recent code from this book's GitHub repository: `https://github.com/PacktPublishing/Computer-Vision-on-AWS`. Clone that repository to your local machine using the following command:

```
$ git clone https://github.com/PacktPublishing/Computer-Vision-
on-AWS
$ cd Computer-Vision-on-AWS/09_SageMakerGroundTruth
```

Additionally, you will need an AWS account and the Jupyter Notebook. `Chapter 1` contains detailed instructions for configuring the developer environment.

Introducing Amazon SageMaker Ground Truth

Amazon SageMaker Ground Truth is a data labeling service. It allows you to generate high-quality training datasets for machine learning models. The service offers two options: Amazon SageMaker Ground Truth Plus and Amazon SageMaker Ground Truth. SageMaker Ground Truth Plus is a turnkey solution where the service takes care of everything from building labeling applications to providing and managing expert workforces so all you need to do is to provide your data along with labeling requirements. We will focus on SageMaker Ground Truth, where you can build and manage your own labeling workflows and workforces. Although not the focus of this chapter, SageMaker Ground Truth can also help you build labeled synthetic datasets – critical for use cases where acquiring real-world data is time-consuming and expensive.

Benefits of Amazon SageMaker Ground Truth

SageMaker Ground Truth provides a number of benefits when it comes to data labeling:

- **Improved data quality and increased visibility**: SageMaker Ground Truth supports a number of features including an automated data labeling capability using active learning to reduce manual burden and errors. It provides logs and metrics of labeling workflows, which increases the visibility of data labeling activities and allows you to track the throughput and efficiency of the labeling workforce.

- **Choose your labeling workforce**: SageMaker Ground Truth provides options to work with internal and external labelers. For example, you can use your own team members to label images, you can access third-party labeling service providers through AWS Marketplace, or you can use a crowdsourcing marketplace through Amazon **Mechanical Turk (MTurk)**.

- **Support for multiple data types**: SageMaker Ground Truth supports labeling multiple data types such as images, text, videos and video frames, and 3D point clouds (3D visualizations of a set of georeferenced points – imagine a 3D visualization of your house).

SageMaker Ground Truth provides several built-in templates to handle different task types, including image classification, object detection (bounding box), semantic segmentation, text classification, **named entity recognition (NER)**, video classification, and video frame object detection. In this chapter, we will use SageMaker Ground Truth to label images and specifically classify and localize objects within an image, using private (your own) labelers.

Automated data labeling

If you need to label large datasets such as thousands of images, the cost and time taken to label these images using only humans would be very high. That's where automated data labeling comes in handy. SageMaker Ground Truth supports active learning using supervised models to automate the labeling of input data. SageMaker Ground Truth supports automated data labeling for task types such as image classification, object detection, semantic segmentation (identifying the content of an image at the pixel level), and text classification.

At a high level, this is how automated data labeling works:

1. You create a labeling job with SageMaker GT and enable it.

2. SageMaker GT randomly samples the dataset and sends it to human workers to label. This step is key as it needs the labeled data to create a training and validation dataset.

3. SageMaker GT uses the labeled training dataset to train the model to be used for auto-labeling.

4. Using the trained model, SageMaker GT will run a batch transform job on the validation dataset and produce a confidence score.

5. Based on the inference results from the validation dataset, GT will derive a confidence threshold that will help determine quality labels.

6. GT will run a batch transform job on the remaining unlabeled dataset and derive a confidence score for each object. Using the confidence threshold from the earlier step, it will determine whether the confidence score for the dataset meets or exceeds the threshold.

7. Objects with a high confidence score are considered auto-labeled.

8. For low confidence scores, GT will send the dataset to human workers. It will then use the newly labeled dataset from human workers to update the model.

9. This process is repeated until the entire dataset is completely labeled (with a high confidence score).

SageMaker GT expects at least 1,250 objects (for example, images) for automated data labeling. However, to achieve higher accuracy for auto-labeling jobs, it is recommended to provide a minimum of 5,000 objects. We will not use the automated data labeling capability for this chapter as we will only be labeling 10 images.

Labeling Packt logos in images using Amazon SageMaker Ground Truth

In this section, we will learn how to create a labeling job to label Packt logo images. We will use a built-in task type (bounding box), will use Amazon S3 to place the input dataset, and will select a private labeling workforce.

Step 1 – collect your images

Upload the sample images from the book's GitHub repository. You can complete this step using the following command:

```
$ aws s3 sync 09_SageMaker_Ground_Truth/images s3://cv-on-aws-
book-xxxx/chapter_09/images --region us-east-2
```

> **Important note**
> To collect the sample images, you can use the same S3 bucket you created in Chapter 2.

Now, navigate to Amazon Lookout for Vision on the AWS Management Console (https://us-east-2.console.aws.amazon.com/sagemaker/home?region=us-east-2#/landing).

Select on **Ground Truth**.

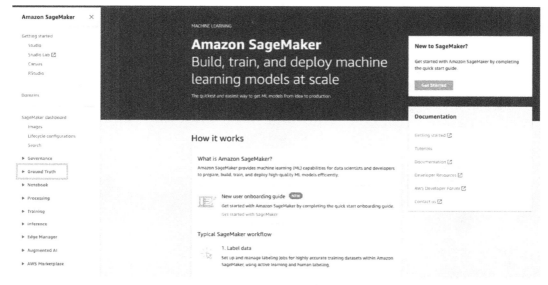

Figure 9.1: Amazon SageMaker Console

> **Important note**
> Your work team, input manifest file, output bucket, and other resources in Amazon S3 must be in the same AWS Region you use to create your labeling job.

Step 2 – create a labeling job

Next, select on **Create labeling job**.

Figure 9.2: Creating a labeling job using the SageMaker console

Give it a name, such as `packt-logo-labeling`.

Step 3 – specify the job details

Next, we need to provide some details about the job, such as the input data setup (the location for input and output datasets), the type of data, and the IAM role for SageMaker to access the data.

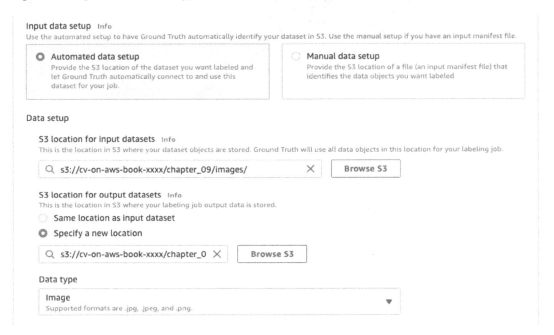

Figure 9.3: Input data setup

The **Input data setup** screen allows you to connect Ground Truth with your input dataset. You can select **Manual data setup** if you have an input manifest file in an S3 bucket. Alternatively, you can use **Automated data setup** if you want Ground Truth to automatically build manifest files. We will use **Automated data setup**.

Next for data setup, we will browse S3 to select locations for input (`s3://cv-on-aws-book-xxxx/chapter_09/images/`) and output datasets (`s3://cv-on-aws-book-xxxx/chapter_09/output_data/`). We will select **Images** as the data type.

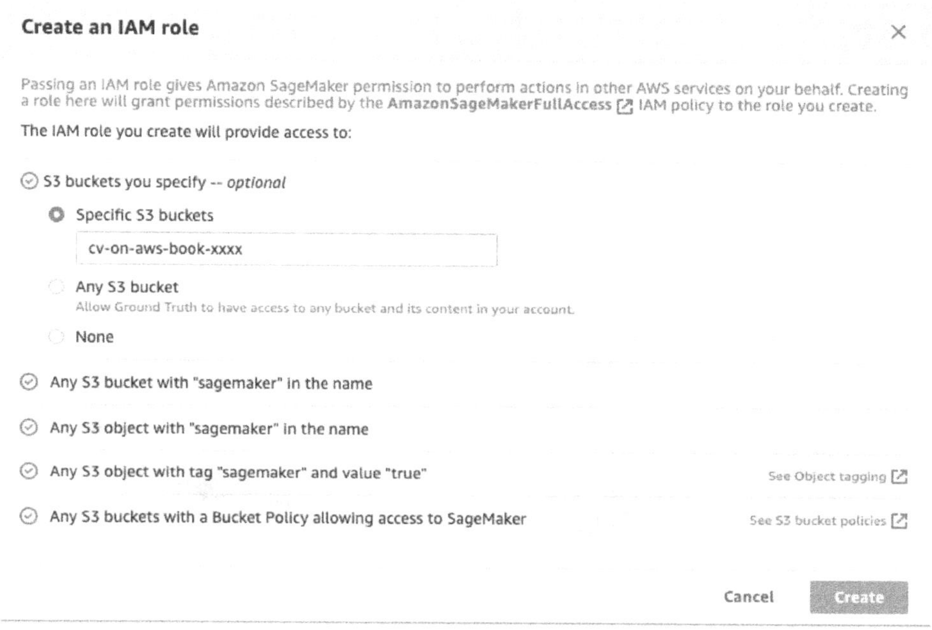

Create an IAM role   ✕

Passing an IAM role gives Amazon SageMaker permission to perform actions in other AWS services on your behalf. Creating a role here will grant permissions described by the **AmazonSageMakerFullAccess** ⬈ IAM policy to the role you create.

The IAM role you create will provide access to:

⊘ S3 buckets you specify -- *optional*

    ⦿ Specific S3 buckets

       | cv-on-aws-book-xxxx |

    ◯ Any S3 bucket
       Allow Ground Truth to have access to any bucket and its content in your account.

    ◯ None

⊘ Any S3 bucket with "sagemaker" in the name

⊘ Any S3 object with "sagemaker" in the name

⊘ Any S3 object with tag "sagemaker" and value "true"   See Object tagging ⬈

⊘ Any S3 buckets with a Bucket Policy allowing access to SageMaker   See S3 bucket policies ⬈

                                             Cancel   `Create`

Figure 9.4: Creating an IAM role

Next, we will create a new IAM role to allow SageMaker to access our S3 bucket. From the dropdown, select on **Create a new role**, provide the S3 bucket name (s3://cv-on-aws-book-xxxx) for input and output datasets, and select **Create**.

Next, select on **Complete data setup**. It may take some time before you see **Input data connection successful** as this process creates an input manifest file on your behalf.

Once we have set up the input data configuration, we will choose the task type in the next step. As we mentioned earlier, there are many built-in task templates available. Otherwise, you have the ability to build your own. Our input data is images. We will select **Image** under **Task category**. As we want workers to draw a bounding box around the Packt logo in our images, we will select **Bounding box** under **Task selection**.

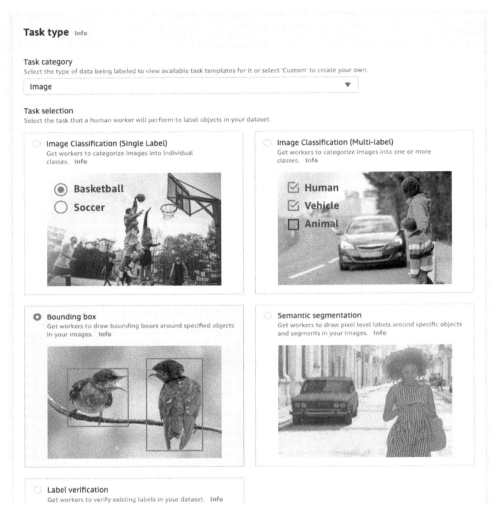

Figure 9.5: Task type configuration

Select **Next**.

Step 4 – specify worker details

In this step, we will configure details about the type of workforce along with some task-specific values.

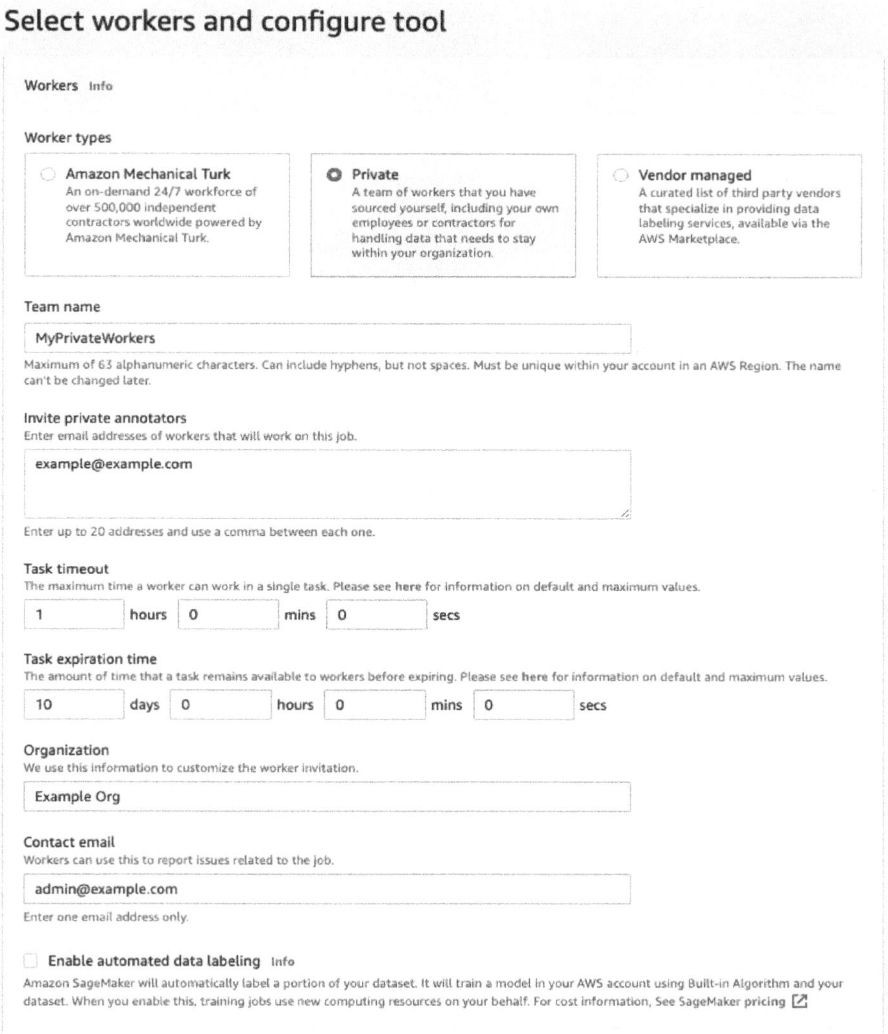

Select workers and configure tool

Workers Info

Worker types

| ⦾ **Amazon Mechanical Turk** | ⦿ **Private** | ⦾ **Vendor managed** |
| An on-demand 24/7 workforce of over 500,000 independent contractors worldwide powered by Amazon Mechanical Turk. | A team of workers that you have sourced yourself, including your own employees or contractors for handling data that needs to stay within your organization. | A curated list of third party vendors that specialize in providing data labeling services, available via the AWS Marketplace. |

Team name

| MyPrivateWorkers |

Maximum of 63 alphanumeric characters. Can include hyphens, but not spaces. Must be unique within your account in an AWS Region. The name can't be changed later.

Invite private annotators
Enter email addresses of workers that will work on this job.

| example@example.com |

Enter up to 20 addresses and use a comma between each one.

Task timeout
The maximum time a worker can work in a single task. Please see here for information on default and maximum values.

| 1 | hours | 0 | mins | 0 | secs |

Task expiration time
The amount of time that a task remains available to workers before expiring. Please see here for information on default and maximum values.

| 10 | days | 0 | hours | 0 | mins | 0 | secs |

Organization
We use this information to customize the worker invitation.

| Example Org |

Contact email
Workers can use this to report issues related to the job.

| admin@example.com |

Enter one email address only.

☐ Enable automated data labeling Info
Amazon SageMaker will automatically label a portion of your dataset. It will train a model in your AWS account using Built-in Algorithm and your dataset. When you enable this, training jobs use new computing resources on your behalf. For cost information, See SageMaker pricing 🗗

Figure 9.6: Worker configuration

For worker types, we have three options: **Public** (Mechanical Turk workforce), **Private** (in-house workforce), and **Vendor** (Marketplace/third-party workforce). For our scenario, we will use private worker types. We will provide a team name of MyPrivateWorkers and add the email addresses of workers. Next, you will provide organization information to customize the invitation email that goes out to workers, along with a contact email to inform workers who they can contact to report labeling-related issues.

We left a few fields, such as **Task timeout** and **Task expiration time**, as the default values and left the automated data labeling box unchecked.

Step 5 – providing labeling instructions

In this step, we will provide labeling instructions that our private workforce will use to label the images. As we are using a built-in task type (bounding box), it comes with a template and we just have to do minor changes such as providing a brief description of the task and adding labels. You can also provide additional instructions or examples of good and bad labeling.

Figure 9.7: Label images

Select on **Create** and you should see the `packt-logo-labeling` job in progress. This status indicates that our labeling job is now available and there are 10 images to be worked upon. Next, we will log in to the labeling portal to start labeling.

Figure 9.8: Labelling job status

Step 6 – start labeling

If you entered your email address in the list of workers/annotators, you should have received an automated email invitation to work on the labeling project we just created.

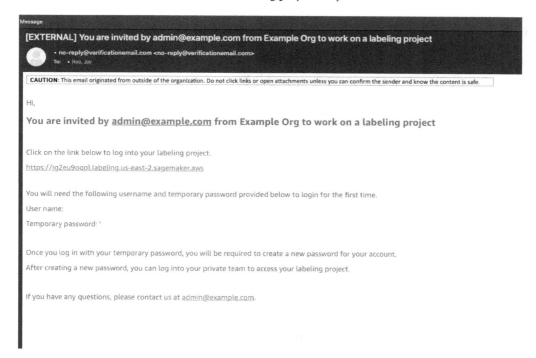

Figure 9.9: Invitation to work on the labeling project

Log in to the worker portal using the email address and temporary password. You'll be prompted to create a new password.

Figure 9.10: Worker portal

Select on **Start working**. You will be taken to a labeling portal where you can see the task description we defined at the top, instructions on the left side, labels to choose from on the right side. and labeling tools at the bottom.

Figure 9.11: Labeling portal

Next, draw a bounding box around the Packt logo and Select **Submit**. Since we only defined a single label, it will auto-select the label.

Figure 9.12: Drawing a bounding box

Continue to label and draw bounding boxes in all the images we submitted. You can do this in one go or you can stop and resume the task later. As a worker, you also have ability to decline if the instructions are not clear. You can also release tasks so they can be assigned to next worker.

While the workers are working on labeling the data, as an administrator, you can observe a labeling summary (number of objects labeled, etc.) in the SageMaker console.

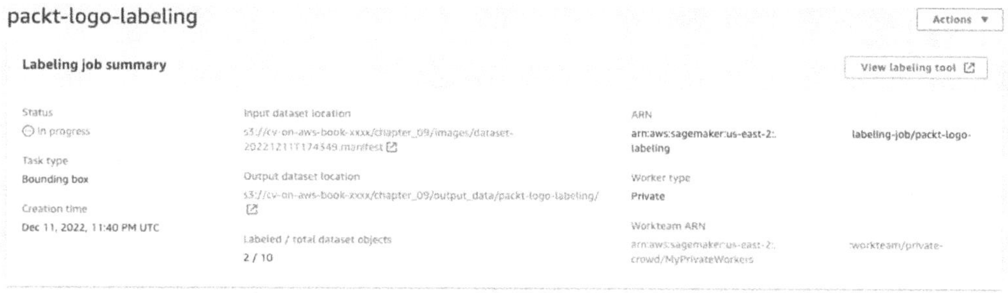

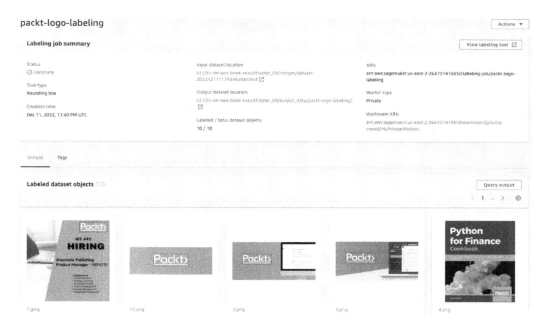

Figure 9.13: Labeling job summary

Once all the objects submitted for labeling under the job are completed, workers will see a message reading **You're finished with the available tasks** in their worker portal. In the SageMaker console, you will see a job summary status of **Complete**. It will also show labeled dataset objects and a bounding box around them.

Figure 9.14: Labeling job summary (2)

Step 7 – output data

Now, we have our labeled dataset, output data is placed in the S3 location you specified while creating the job. Within the output location, you will see a subfolder for each job submitted, a folder named `packt-logo-labeling` in our case.

There are a few subfolders at this location:

- `annotation-tool`: This contains information about all the labels created in this job along with the template we defined. The template type is liquid (an open source template language).

- `annotations`: This contains few subdirectories where you can find responses from individual workers as well as the consolidation/combination of annotations of two or more workers. Ground Truth provides a built-in annotation consolidation function for multiple labeling tasks, including bounding boxes.

- `manifests`: This is the key folder where service stores output the labeling job. Each entry in the output data file is identical to the input manifest file along with an attribute and value for the label assigned to the input object.

```
output.manifest
1  {"source-ref":"s3://cv-on-aws-book-xxxx/chapter_09/images/1.jpeg","packt-l
   ogo-labeling":{"image_size":[{"width":800,"height":800,"depth":3}],"annota
   tions":[{"class_id":0,"top":19,"left":512,"height":93,"width":265}]},"pack
   t-logo-labeling-metadata":{"objects":[{"confidence":0}],"class-map":{"0":"
   packt"},"type":"groundtruth/object-detection","human-annotated":"yes","cre
   ation-date":"2022-12-12T00:49:21.818147","job-name":"labeling-job/
   packt-logo-labeling"}}
2  {"source-ref":"s3://cv-on-aws-book-xxxx/chapter_09/images/10.png","packt-l
   ogo-labeling":{"image_size":[{"width":277,"height":277,"depth":3}],"annota
   tions":[{"class_id":0,"top":69,"left":237,"height":121,"width":347}]},"pac
   kt-logo-labeling-metadata":{"objects":[{"confidence":0}],"class-map":{"0":
   "packt"},"type":"groundtruth/object-detection","human-annotated":"yes","cr
   eation-date":"2022-12-12T00:48:16.819349","job-name":"labeling-job/
   packt-logo-labeling"}}
3  {"source-ref":"s3://cv-on-aws-book-xxxx/chapter_09/images/2.png","packt-lo
   go-labeling":{"image_size":[{"width":828,"height":386,"depth":3}],"annotat
   ions":[{"class_id":0,"top":122,"left":98,"height":102,"width":292}]},"pack
   t-logo-labeling-metadata":{"objects":[{"confidence":0}],"class-map":{"0":"
   packt"},"type":"groundtruth/object-detection","human-annotated":"yes","cre
   ation-date":"2022-12-12T00:49:21.821566","job-name":"labeling-job/
   packt-logo-labeling"}}
```

Figure 9.15: Output manifest file

- `temp` – A temporary location for service use.

Now that we know how to work with Ground Truth to label data, we will get started with training models in the next section.

Importing the labeled data with Rekognition Custom Labels

Now we have labeled data, we can use it to train machine learning models with SageMaker or Rekognition Custom Labels. We will navigate to the Rekognition Custom Labels console (https://us-east-2.console.aws.amazon.com/rekognition/custom-labels#/).

Step 1 – create the project

Navigate to the **Projects** panel on the left sidebar and select on **Create project**. Give it a name, such as `logo-detection`.

Step 2 – create training and test datasets

Next, we need to create a dataset. Select on **Create dataset**. Select **Start with a single dataset** and then select **Import images labeled by SageMaker Ground Truth** under **Training dataset details**.

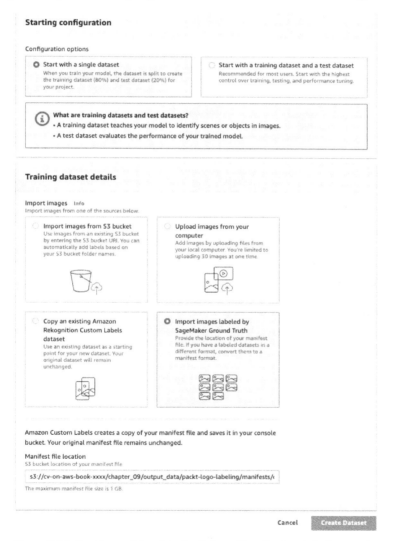

Figure 9.16: Creating a dataset in the Rekognition Custom Labels console

Provide the location of output manifest file, such as `s3://cv-on-aws-book-xxxx/chapter_09/output_data/packt-logo-labeling/manifests/output/output.manifest`, and select **Create Dataset**.

As you can see, the labeled images were imported into the project with all the labels and bounding box information.

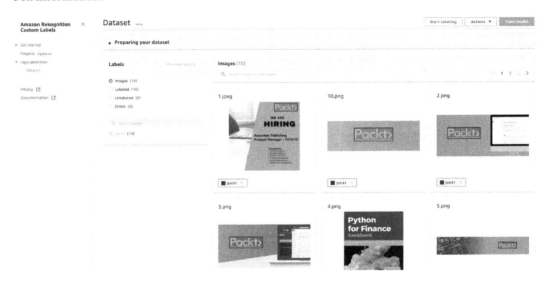

Figure 9.17: Dataset in the Rekognition Custom Labels console

Step 3 – model training

If you have a need, you can further edit the bounding box or labels in the Rekognition Custom Labels console. If not, you can select on **Train model** to use the dataset to start the model training.

Figure 9.18: Model training complete

Once the model training finishes, you can start using the model similar to how we learned in *Chapter 3*. Essentially, there is no difference in the Rekognition Custom Labels model training process whether the images were labeled in the console or with SageMaker Ground Truth and later imported into Custom Labels. Similarly, you can also import the SageMaker Ground Truth labeled images into Lookout for Vision projects.

Summary

In this chapter, we covered what SageMaker Ground Truth is and how you can use it to label datasets. We discussed how you can create labeling jobs to label images, text, video, and 3D point clouds for various task types. We covered how you can use one of the built-in task types, bounding box, a labeling task to classify and localize the Packt logo within images.

Now that you have learned how to label datasets, in the next chapter you will learn how you can build computer **CV models** using **Amazon SageMaker**.

10

Using Amazon SageMaker for Computer Vision

Amazon Rekognition is a purpose-built computer vision service for 80% of everyday use cases. But what if you need more control and influence over that remaining 20%? In that case, you'd choose tooling from a lower level of AWS AI Layered Cake. These situations are precisely where Amazon SageMaker shines:

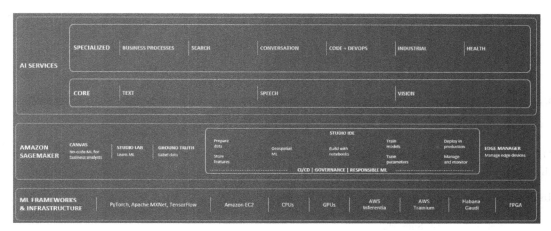

Figure 10.1 – AWS AI Layered Cake

Amazon SageMaker aims to make **machine learning** (**ML**) possible for every developer, business analyst, and data scientist. It achieves this goal through a fully managed suite of services and tools that address each ML model life cycle management step. You can start building with fully managed Juypter notebooks and train your model at scale using elastic resource pools. Suppose you need features for data labeling, data wrangling, and creating **human-in-the-loop** (**HITL**) workflows. In that case, Amazon SageMaker GroundTruth, Data Wrangler, and **Augmented AI** (**A2I**) have you covered.

You can choose the subset of functionality necessary to build, train, infer, govern, and monitor ML solutions in the most demanding production environments. Nearly 80% of ML production costs occur after creating the model. That's why Amazon SageMaker includes capabilities for drift detection, governance, hosting inference, and operating at the edge. You can also monetize custom models through AWS Marketplace:

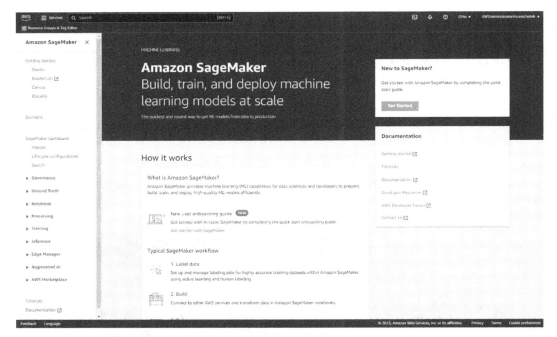

Figure 10.2 – The Amazon SageMaker homepage

You could fill an entire bookshelf detailing SageMaker's capabilities. So, in this chapter, you'll only scratch the surface by recreating the famous Hotdog, Not-Hotdog classifier using a built-in algorithm.

In this chapter, we will cover the following key topics:

- What is Amazon SageMaker?
- Using the built-in image classifier
- Troubleshooting the training job
- Handling binary metadata files

Technical requirements

A Jupyter notebook is available for running the example code from this chapter. You can access the most recent code from this book's GitHub repository at `https://github.com/PacktPublishing/Computer-Vision-on-AWS`.

Clone that repository to your local machine using the following command:

```
$ git clone https://github.com/PacktPublishing/Computer-Vision-
on-AWS
$ cd Computer-Vision-on-AWS/10_SageMakerModel
```

Additionally, you will need an AWS account and Jupyter Notebook. *Chapter 1,* contains detailed instructions for configuring the development environment.

Fetching the LabelMe-12 dataset

Rafael Uetz and Sven Behnke's `LabelMe-12` dataset contains 50,000 labeled images from twelve categories. You'll use it later in the chapter to learn how to import datasets that use non-standard file structures and binary metadata. See `https://www.ais.uni-bonn.de/download/datasets.html` for more information about this open source artifact.

Installing prerequisites

You can execute shell commands from the Jyupter notebook directly. Create a code block using the bang (!) command syntax. This handy approach lets you avoid needing a separate terminal window:

```
!pip3 install requests
```

If you are using the terminal screen, simply remove the bang prefix:

```
$ pip3 install requests
```

The command outputs the following report:

```
Collecting requests
  Downloading requests-2.28.2-py3-none-any.whl (62 kB)
     |████████████████████████████| 62 kB 665 kB/s eta
0:00:01
Installing collected packages: urllib3, idna, charset-
normalizer, certifi, requests
Successfully installed certifi-2022.12.7 charset-
normalizer-3.0.1 idna-3.4 requests-2.28.2 urllib3-1.26.14
```

Downloading the tar.gz file

Next, download the `LabelMe-12-50k.tar.gz` file into the project filesystem:

```
import requests
import os
download_file = 'https://www.ais.uni-bonn.de/deep_learning/
LabelMe-12-50k.tar.gz'
local_file = os.path.join('raw',os.path.basename(download_
file))
if not os.path.exists('raw'):
    os.mkdir('raw')


r = requests.get(download_file)
with open(local_file, 'wb') as f:
    f.write(r.content)
```

Extracting the tar.gz file

Next, you'll need to decompress the dataset and the `extractall` files from the tarball. The files will expand into the `./data/test` and `./data/train` folders:

```
import tarfile
with tarfile.open(local_file) as f:
    f.extractall('raw')
```

Installing TensorFlow 2.0

Use the following command to install TensorFlow into your SageMaker notebook:

```
!pip3 install tensorflow
```

You will see an output similar to the following:

```
Looking in indexes: https://pypi.org/simple, https://pip.repos.
neuron.amazonaws.com
Collecting tensorflow
  Downloading tensorflow-2.11.0-cp310-cp310-manylinux_2_17_
x86_64.manylinux2014_x86_64.whl (588.3 MB)
                                    588.3/588.3 MB 1.9 MB/s
eta 0:00:0000:0100:01
```

```
Requirement already satisfied: h5py>=2.9.0 in /home/ec2-user/
anaconda3/envs/python3/lib/python3.10/site-packages (from
tensorflow) (3.7.0)

...

Successfully installed abs1-py-1.4.0 astunparse-1.6.3
cachetools-5.3.0 flatbuffers-23.1.21 gast-0.4.0 google-
auth-2.16.0 google-auth-oauthlib-0.4.6 grpcio-1.51.1
keras-2.11.0 libclang-15.0.6.1 markdown-3.4.1 oauthlib-3.2.2
opt-einsum-3.3.0 protobuf-3.19.6 pyasn1-modules-0.2.8
requests-oauthlib-1.3.1 tensorboard-2.11.2 tensorboard-data-
server-0.6.1 tensorboard-plugin-wit-1.8.1 tensorflow-2.11.0
tensorflow-estimator-2.11.0 tensorflow-io-gcs-filesystem-0.30.0
termcolor-2.2.0
```

Confirm that TensorFlow imports successfully:

```
Import tensorflow
```

This command will output the following initialization report. Suppose you're running an instance without GPU support. In that case, the warning will state that NVIDIA GPU support (CUDA) is missing:

```
2023-02-05 00:46:55.973149: I tensorflow/core/platform/cpu_
feature_guard.cc:193] This TensorFlow binary is optimized
with oneAPI Deep Neural Network Library (oneDNN) to use
the following CPU instructions in performance-critical
operations:  AVX2 AVX512F FMA
To enable them in other operations, rebuild TensorFlow with the
appropriate compiler flags.
2023-02-05 00:46:56.217457: W tensorflow/compiler/xla/stream_
executor/platform/default/dso_loader.cc:64] Could not load
dynamic library 'libcudart.so.11.0'; dlerror: libcudart.
so.11.0: cannot open shared object file: No such file or
directory
2023-02-05 00:46:56.217478: I tensorflow/compiler/xla/stream_
executor/cuda/cudart_stub.cc:29] Ignore above cudart dlerror if
you do not have a GPU set up on your machine.
2023-02-05 00:46:57.703897: W tensorflow/compiler/xla/stream_
executor/platform/default/dso_loader.cc:64] Could not load
dynamic library 'libnvinfer.so.7'; dlerror: libnvinfer.so.7:
cannot open shared object file: No such file or directory
2023-02-05 00:46:57.703981: W tensorflow/compiler/xla/stream_
executor/platform/default/dso_loader.cc:64] Could not load
dynamic library 'libnvinfer_plugin.so.7'; dlerror: libnvinfer_
```

```
plugin.so.7: cannot open shared object file: No such file or
directory
2023-02-05 00:46:57.703990: W tensorflow/compiler/tf2tensorrt/
utils/py_utils.cc:38] TF-TRT Warning: Cannot dlopen some
TensorRT libraries. If you would like to use Nvidia GPU with
TensorRT, please make sure the missing libraries mentioned
above are installed properly.
```

Suppose you rerun this command on a GPU instance such as `ml.g4dn.xlarge`. In that case, you should expect the following output:

```
2023-02-05 02:11:30.623259: I tensorflow/core/platform/cpu_
feature_guard.cc:193] This TensorFlow binary is optimized
with oneAPI Deep Neural Network Library (oneDNN) to use
the following CPU instructions in performance-critical
operations:  AVX2 AVX512F AVX512_VNNI FMA
To enable them in other operations, rebuild TensorFlow with the
appropriate compiler flags.
2023-02-05 02:11:30.743717: I tensorflow/core/util/port.cc:104]
oneDNN custom operations are on. You may see slightly different
numerical results due to floating-point round-off errors
from different computation orders. To turn them off, set the
environment variable `TF_ENABLE_ONEDNN_OPTS=0`.
```

Installing matplotlib

The `matplotlib` Python plotting library provides an object-oriented API for embedding plots into applications. You'll use it for rendering images into the Jupyter notebook:

```
!pip3 install matplotlib
```

Using the built-in image classifier

Amazon SageMaker's built-in algorithms and pre-trained models address everyday use cases such as image classification, text summarization, and anomaly detection. In this section, you'll recreate the famous Hot Dog, Not-Hot Dog classifier using MXNet by following these steps:

1. Upload the dataset to Amazon S3.
2. Create the training job definition.
3. Run the job.
4. Verify the results.

Upload the dataset to Amazon S3

This chapter's repository includes the hotdog-nothotdog image set. Upload these files into an Amazon S3 bucket:

```
$ aws s3 sync 10_SageMakerModel/hotdog-nothotdog s3://ch10-cv-
book-use2/sagemaker/hotdog-nothotdog --region us-east-2
```

Prepare the job channels

Amazon SageMaker uses channels to sequence the training, testing, and validation data into the models and algorithms. The channel's format will vary from simple delineated manifest files (e.g., using tabs or commas) to RecordIO pipelines.

For this task, you must create a tab-separated file with columns for the image's identifier, label, and relative location. Producing these files is straightforward, using a few lines of Python code and the Amazon S3 boto3 module: The following steps will show you how:

1. Initialize the Amazon S3 client.
2. Enumerate the bucket's objects.
3. Sort the objects into the test and train datasets.
4. Create the channel files.
5. Start the training job.
6. Wait for completion.

Initialize the boto3 client

This snippet declares the location of our test files and initializes s3_client. You'll need to configure bucket_name and region_name to match your environment:

```
bucket_name='ch10-cv-book-use2'
prefix = 'sagemaker/hotdog-nothotdog'
region_name='us-east-2'

import boto3
s3_client = boto3.client('s3',region_name=region_name)
```

Enumerate the bucket's objects

Next, let's fetch the dataset's file names from the S3 bucket. The recommended method for enumerating those objects is with the ListObjectsV2 API. As the name suggests, the ListObjectsV2 API returns up to 1,000 objects that match a given filter.

Like other APIs, you can use the response's NextToken property to retrieve the next page's results iteratively. You have already seen in previous chapters how to write this tedious code. Instead, let's use the boto3 client's get_paginator method to create the PageIterator object:

```
files=[]
paginator = s3_client.get_paginator('list_objects_v2')
pages = paginator.paginate(Bucket=bucket_name, Prefix=prefix)
for page in pages:
    for obj in page['Contents']:
        files.append(obj['Key'])
```

Sort the objects into datasets

The files list contains the recursive object keys below our S3 prefix in the sagemaker/hotdog-nothotdog/DATASET_NAME/LABEL/file.jpg format:

- DATASET_NAME is train or test
- LABEL is hotdog or nothotdog

Let's parse the folder structure into the respective dataset components:

```
def get_dataset(name):
    ds = {'hotdog':[], 'nothotdog':[]}
    for file in files:
        if not file.endswith('.jpg'):
            continue
        if not '/%s/' % name in file:
            continue
        if '/hotdog/' in file:
            ds['hotdog'].append(file)
        elif '/nothotdog/' in file:
            ds['nothotdog'].append(file)
    return ds
```

Finally, call the function once per dataset to populate the train_ds and test_ds dictionaries:

```
train_ds = get_dataset('train')
test_ds = get_dataset('test')
```

Create the channel files

Amazon SageMaker's image classification algorithm needs two manifest files that specify which S3 objects to include in the train and test (validation) workstreams.

Let's construct the `train_lst` and `validation_lst` manifest files using the results of the `get_dataset` function. These files must be tab-separated with columns for the `identifier`, `label`, and `relative path`:

Column Name	Format	Description
identifier	Number	The file index or name for troubleshooting purposes
label	Number	A numeric value starting at zero that represents the file's label (0=nothotdog, 1=hotdog)
relative path	String	The relative path to the dataset's root directory (label/identifier.jpg)

Table 10.1 – The channel column names

Use the following code to transform the `get_dataset` function's result into the expected format:

```
from os import path
def create_channel(ds):
    channel=[]
    for label in ds.keys():
        for obj in ds[label]:
            identifier = path.splitext(path.basename(obj))[0]
            relpath = '%s/%s.jpg' % (label, identifier)
            class_id = 1 if label == 'hotdog' else 0
            channel.append('%s\t%s\t%s' %(
                identifier,
                class_id,
                relpath
            ))
    return channel
```

Upload to the Amazon S3 bucket

Finally, call the `create_channel` function once per data to initialize the `train_lst` and `validation_lst` manifest files. Then upload the resulting strings to the S3 bucket:

```
train_lst = '\n'.join(create_channel(train_ds))
validation_lst = '\n'.join(create_channel(test_ds))

s3_client.put_object(
    Bucket=bucket_name,
    Key='sagemaker/hotdog-nothotdog/train.lst',
    Body=train_lst)

s3_client.put_object(
    Bucket=bucket_name,
    Key='sagemaker/hotdog-nothotdog/validation.lst',
    Body=test_lst)
```

Quick recap

In this task, you retrieved the testing and training files from Amazon S3. Next, you sorted the list into respective lists of `hotdog` and `nothotdog` for those datasets. Finally, you encoded those lists into channel manifest files and persisted them to your S3 bucket. With that out of the way, you can start the training job!

Start the training job

The easiest way to get started with a built-in algorithm or pre-trained model is through the AWS Management Console. After verifying the parameters work as intended, you can export the training job using the `DescribeTrainingJob` API:

1. Navigate to the Amazon SageMaker console `https://us-east-2.console.aws.amazon.com/sagemaker/home?region=us-east-2#/studio`.

2. Expand the **Training** menu and select the **Training jobs** link.

3. Click the **Create training job** button.

Job settings pane

The **Job settings** pane specifies general properties such as **Job name**, **IAM role**, and **Algorithm options**. Specify the built-in algorithm **Vision – Image Classification (MxNet)** from the drop-down menu and leave the remaining defaults:

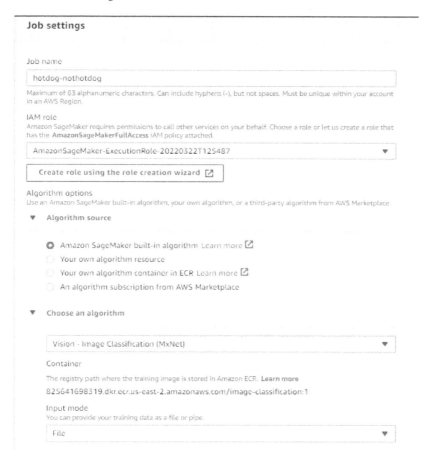

Figure 10.3 – The Job Settings pane

The Resource configuration pane

This particular algorithm requires a GPU for training, such as the `ml.g4dn.xlarge` or `ml.p2.xlarge` instance types.

While experimenting, you should also set the **Additional storage volume per instance (GB)** property to a larger-than-expected value. If the training instance runs out of local storage, the job will crash and terminate the run:

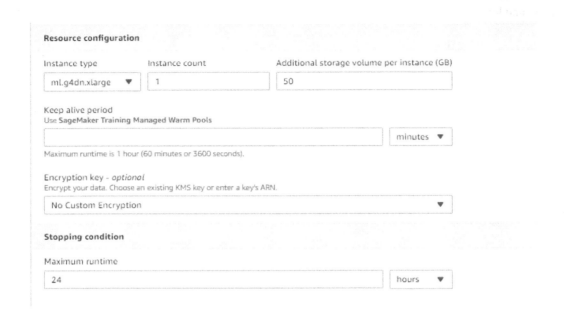

Figure 10.4 – The Resource configuration pane

The Network pane

This built-in algorithm doesn't require a private VPC. You use this option for custom training jobs that need to access protected data sources such as Amazon Redshift, Aurora, and internal APIs:

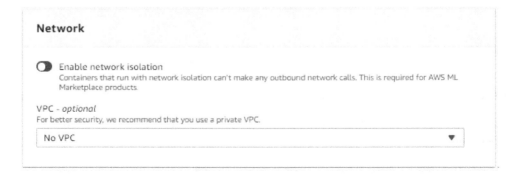

Figure 10.5 – The Network pane

The Hyperparameters pane

An ML model internally contains connected weighted layers of neurons called *parameters*. In contrast, *hyperparameters* are the job arguments that control the training job's behavior.

Configure the following parameters and leave the remaining defaults:

- Set num_classes to 2
- Set num_training_samples to 3000
- Set early_stopping to 1

The following table enumerates the image classification algorithm's hyperparameters and their meaning:

Parameter Name	Required	Description
num_classes	Yes	The total number of labels in the dataset
num_training_samples	Yes	The total training examples for distributing the training dataset
Augmentation_type	No	This modifies the images with crop, reduces colors (crop_color), or completes transform (crop_color_transform) operations
beta_1	No	This is the first exponential decay parameter for the adam optimizer
beta_2	No	This is the second exponential decay parameter for the adam optimizer
checkpoint_frequency	No	The total epochs before saving the model.tar.gz file
early_stopping	No	This enables logic to exit training early if additional epochs (iterations) aren't necessary
early_stop_ping_min_epochs	No	The minimum epochs (iterations) to perform if early_stopping is enabled
early_stopping_patience	No	The total epochs (iterations) of no improvement before the early_stopping logic exits
early_stopping_tolerance	No	This specifies early_stopping based on the accuracy divided by the best accuracy
Epochs	No	The total training iterations
Eps	No	The epsilon for the adam and rmsprop optimizers – used to avoid division by zero

Gamma	No	The decay factor for the `rmsprop` optimizer
`image_shape`	No	This is a three-numeric tuple representing `num_channels`, `height`, and `width` of the images
`kv_store`	No	This specifies whether distributed training shares weights synchronously (`dist_sync`) or asynchronously (`dist_async`)
`learning_rate`	No	The initial learning rate
`lr_scheduler_factor`	No	The ratio to reduce `learning_rate` in conjunction with the `lr_scheduler_step` parameter
`mini_batch_size`	No	The total files to process per batch
Momentum	No	The momentum for the `sgd` and `nag` optimizers
`multi_label`	No	A flag for enabling multiple labels per file
`num_layers`	No	The total layers in the ML model
`optimizer`	No	The optimizer for training the model: • sgd: The stochastic gradient descent • adam: The adaptive momentum estimation • rmsprop: The root mean square propagation • nag: The Nesterov accelerated gradient
`precision_dtype`	No	This specifies whether model weights are single (`float32`) or half (`float16`) precision-values
Resize	No	The image resizing for training data
`top_k`	No	This reports the `top-k` accuracy during training
`use_pretrained_model`	No	A flag to use a pre-trained image network and avoid tuning the initial layers
`use_weighted_loss`	No	A flag to use the weighted cross-entropy loss for the `multi_label` classification
`weight_decay`	No	The coefficient weight decay for the `sgd` and `nag` optimizers

Table 10.2 – The image classification algorithm's hyperparameters

Input data configuration pane

The built-in image classification algorithm needs four resource channels to train the model. In the *Prepare the job channels* section, you already created and uploaded the `train_lst` and `validation_lst` `manifests` into S3.

The following table enumerates the expected channel types for the image classification algorithm:

Channel Name	Type	Description
`train`	S3 prefix	This is the base prefix of the training files
`train_lst`	Tab-separated manifest file	This is a three-column manifest listing of training files to include
`validation`	S3 prefix	This is the base prefix of the validation files
`validation_ lst`	Tab-separated manifest file	This is a three-column manifest listing of validation files to include

Table 10.3 – Required channels

The S3 prefix channels

Add two channels named `train` and `validation` with the default values intact:

- Set the **S3 data type** field to **S3Prefix**
- Set the **S3 location** field to `s3://<yourbucket>/sagemaker/hotdog-nothotdog/CHANNEL_NAME`

Confirm you replaced the CHANNEL_NAME placeholder with train or validation:

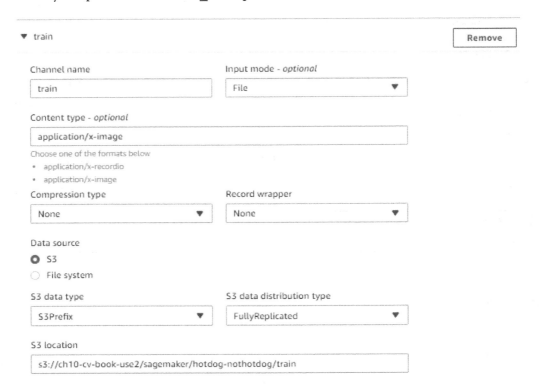

Figure 10.6 – The S3 prefix channels

The manifest channels

Add two channels named train_1st and validation_1st with the default values intact:

- Set the **S3 data type** field to **ManifestFile**
- Set the **S3 location** field to s3://<yourbucket>/sagemaker/hotdog-nothotdog/CHANNEL_NAME.1st

Confirm you replaced the CHANNEL_NAME placeholder with train or validation:

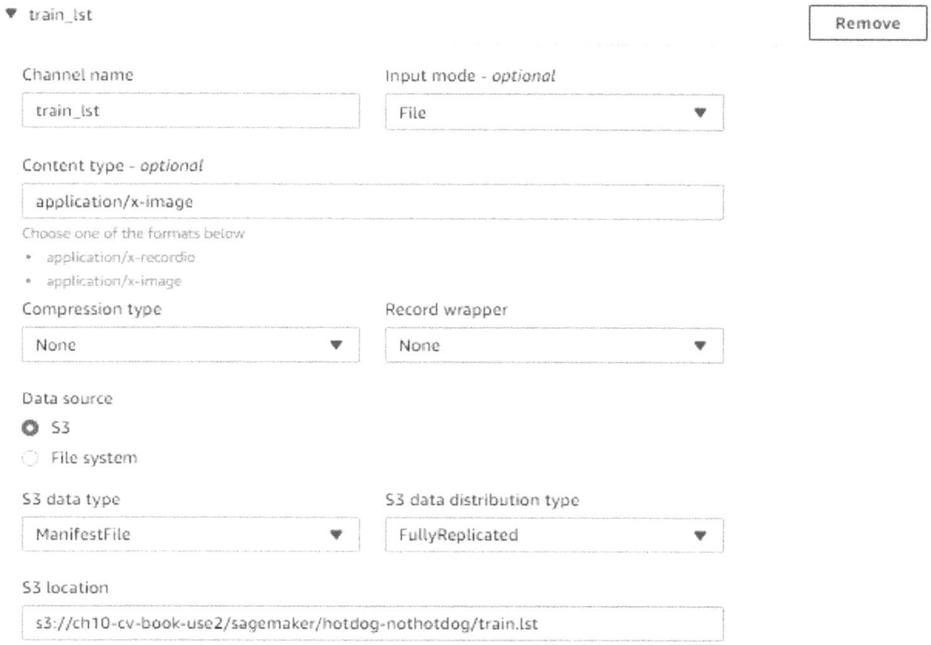

Figure 10.7 – The manifest file channel

The augmented manifest file channel

Instead of specifying the tab-delimited ManifestFile data type, you can optionally use a JSON-encoded AugmentedManifest data type. These files require a complete JSON structure per line:

```
{"source-ref":"s3://bucket/hotdog1.jpg", "class":"1"}
{"source-ref":"s3://bucket/nothotdog2.jpg", "class":"0"}
{"source-ref":"s3://bucket/hotdog1.jpg", "class":"1"}
```

Alternatively, you can specify one-hot encoding by setting class to a list of zeros and ones. In this case, dog1.jpg and notdog2.jpg have labels of 1 and 0, respectively:

```
{"source-ref":"s3://bucket/dog1.jpg", "class":"[0,1]"}
{"source-ref":"s3://bucket/notdog2.jpg", "class":"[1,0]"}
```

The Checkpoint configuration pane

Optionally, you can turn on checkpointing to serialize the model into an S3 bucket. This feature speeds up restarting failed models and troubleshooting scenarios:

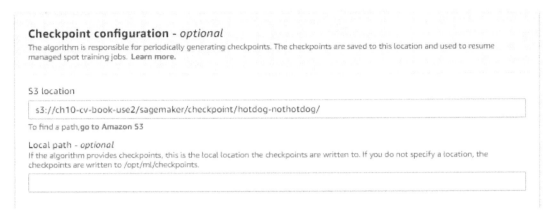

Figure 10.8 – Checkpoint configuration

The Output data configuration pane

Specify in the **S3 output path** field where SageMaker's execution role will write the trained model file.

Optionally, you can choose an **AWS Key Management Service (AWS KMS)** encryption key instead of the default S3 bucket key. This setting helps protect your production models by ensuring that only authorized identities can read the model:

Output data configuration

S3 output path

s3://ch10-cv-book-use2/sagemaker/output/hotdog-nothotdog/

Encryption key - *optional*
If you want Amazon SageMaker to encrypt the output of your training job using your own AWS KMS encryption key instead of the default S3 service key, provide its ID or ARN.

Figure 10.9 – Output data configuration

The Managed spot training pane

Amazon EC2 Spot Instances let you take advantage of the unused EC2 capacity in the AWS cloud. You can save up to 90% over on-demand instances! Suppose AWS needs that capacity back. In that case, the EC2 instance receives a two-minute notification to checkpoint the state and begin termination.

This design is ideal for distributed training, stateless workloads, big data analytics, and containerized applications.

> **Embrace EC2 Spot Instances**
>
> You should be aware of the early termination design, but don't fear it. AWS reclaims less than 5% of EC2 instances, and countless mission-critical production applications use it at scale.

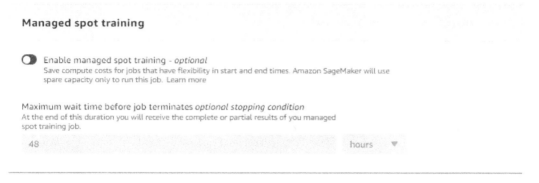

Figure 10.10 – Managed spot training

The Tags pane

Associating tags with your training jobs adds metadata for tracking costs and monitoring auxiliary resource creation. For example, you can tag the job with the caller's team name and cost center. That information appears in AWS Cost Explorer and simplifies departmental chargebacks:

Figure 10.11 – The Tags pane

Submitting the job button

Choose the **Create training job** button when you've finished populating the training job configuration. This action will invoke SageMaker's `CreateTrainingJob` API to provision the compute, sequence the images, and output `model.tar.gz` in S3.

> **This step will take a while**
>
> The training job takes 6,000 seconds using the `1 x ml.p2.xlarge` instance. You can also use `1x ml.g4dn.16xlarge` to finish in 1,827 seconds. Since AWS charges training time by the second, p2 costs $1.50 versus g4dn which costs $2.20.
>
> Alternatively, use this time to rewatch the classic episode of HBO's *Silicon Valley*, *Season 4, Episode 4: Teambuilding Exercise*.

Programmatically creating the job

You can experiment and iterate quickly through the AWS Management Console. After settling on a specific training configuration, you can retrieve it using the AWS **command line interface** (**CLI**). This chapter's repository contains an example as `describe-training-job.json`:

```
$ aws sagemaker describe-training-job --training-job-name
hotdog-nothotdog --region us-east-2
```

Use these values with the Amazon SageMaker's client using `boto3`:

```
import boto3
region_name = "us-east-2"
sagemaker = boto3.client(
    "sagemaker",
    region_name="region_name")

sagemaker.create_training_job(...)
```

Quick recap

In this section, you learned how to pass the channels to the (overly?) flexible `CreateTrainingJob` API. This action will eventually produce the model using the custom data.

Monitoring and troubleshooting

You can view statistics and troubleshooting information when the model has completed training.

View the job history

An exciting view is the **Status history** dialog box, which shows the training job's high-level stages. You can use this information to observe how long different stages take or track down failure reasons:

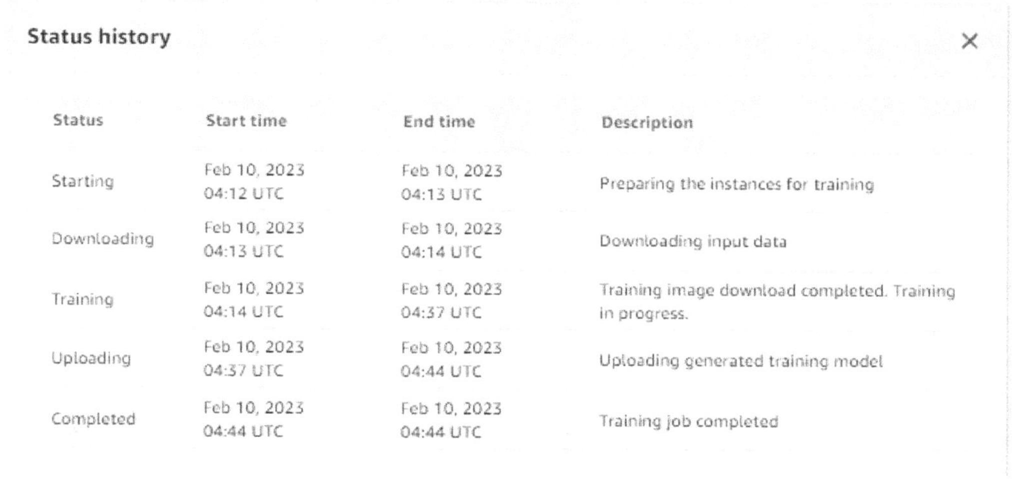

Status history			✕
Status	Start time	End time	Description
Starting	Feb 10, 2023 04:12 UTC	Feb 10, 2023 04:13 UTC	Preparing the instances for training
Downloading	Feb 10, 2023 04:13 UTC	Feb 10, 2023 04:14 UTC	Downloading input data
Training	Feb 10, 2023 04:14 UTC	Feb 10, 2023 04:37 UTC	Training image download completed. Training in progress.
Uploading	Feb 10, 2023 04:37 UTC	Feb 10, 2023 04:44 UTC	Uploading generated training model
Completed	Feb 10, 2023 04:44 UTC	Feb 10, 2023 04:44 UTC	Training job completed

Figure 10.12 – The Status history pane

Monitoring progress

The **Monitor** pane on the job's detail page provides one-click links to Amazon CloudWatch. Reviewing the CloudWatch logs is extremely helpful for understanding why a training job hasn't completed successfully. You can manually find the logs by taking the following steps:

1. Navigate to the CloudWatch console `https://us-east-2.console.aws.amazon.com/cloudwatch/home?region=us-east-2`.
2. Expand **Logs** and choose **Log groups.**
3. Find the **/aws/sagemaker/TrainingJobs** Log group.

4.   Find the **Log stream** that begins with your training job's name:

Figure 10.13 – CloudWatch Logs

The SageMaker algorithm training page also contains nine inline performance graphs to help track down under and over-provisioned resources. This information also resides within CloudWatch metrics' `/aws/sagemaker/TrainingJobs` namespace.

You can use these metrics to scale your resources efficiently. Suppose the job is completed with no or low GPU utilization. In that case, adding more GPU resources won't improve the overall performance. You can also use the validation set's accuracy/epochs to gauge incremental improvements. When this metric becomes flat, it signifies that adding more epochs (training cycles) won't improve the prediction quality:

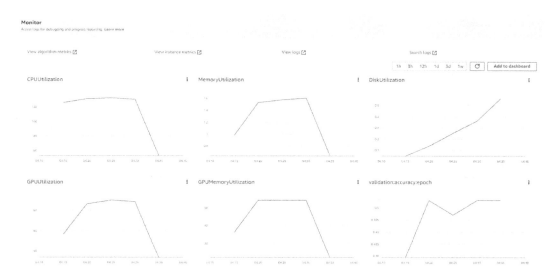

Figure 10.14 – Training job monitoring

The following table enumerates the meaning of these metrics:

Metric Name	Description
CPUUtilization	This is the average CPU utilization across the training instances
MemoryUtilization	This is the average memory utilization across the training instances
DiskUtilization	This is the average disk space utilization across the training instances
GPUUtilization	This is the average GPU compute utilization across the training instances
GPUMemoryUtilization	This is the average GPU memory utilization across the training instances
validation:accuracy:epoch	This is the validation dataset's accuracy improvement per epoch
validation:accuracy	This is the validation dataset's accuracy
train:accuracy:epoch	This is the training dataset's accuracy improvement per epoch
train:accuracy	This is the training dataset's accuracy

Table 10.4 – Default training metrics

Quick recap

In this section, you learned how to use the built-in image classification algorithm to solve the hotdog-nothotdog problem. It might have seemed like a lot of effort to make something so trivial, but rest assured, you got more out of it. This same process applies to customizing virtually any built-in or AWS Marketplace algorithm! Now you can tap into hundreds of custom AI capabilities for your applications.

Next up, what if you need to train with open source code that requires a bit of finesse to prepare?

Handling binary metadata files

Ingesting the hotdog-nothotdog dataset was pretty straightforward since you only needed to parse the path. However, many scientific and open source datasets encode their labels into binary files. This approach reduces the data size and increases file parsing performance. You're probably wondering whether we can still use them. Of course!

The LabelMe-12 dataset from the technical requirements section is one such example. It includes the label information in the annotation.bin (binary) and annotation.txt (human-readable) files under the ./data/test and ./data/train folders. Let's focus on the binary file and only use the human-readable copy for troubleshooting.

We will perform the following steps to do this:

1. Declare the custom Label enumeration.

2. Declare the custom Annotation class.

3. Read each image's labels from the file.

4. Confirm the expected counts are present.

5. Normalize the file structure.

Declaring the Label class

The LabelMe-12 dataset contains 50,000 images that align to twelve everyday object types. Let's start by declaring a Label class with these labels as constants so we can make our values developer friendly:

```
from enum import Enum
class Label(Enum):
    NONE=0
    PERSON=1
    CAR=2
    BUILDING=3
    WINDOW=4
```

```
TREE=5
SIGN=6
DOOR=7
BOOKSHELF=8
CHAIR=9
TABLE=10
KEYBOARD=11
HEAD=12
CLUTTER= 13
```

Reading the annotations file

Uetz and Behnke serialized the label values using one-hot encoding with twelve 32-bit float point values per image annotation. Let's reflect for a moment on what that sentence means.

First, the image's annotation uses one-hot encoding, so we know only one of the twelve values will be `true` (1). Therefore, whichever one-hot encoded `index` equals is true will equal the label's index.

> **Important note**
> Python arrays use a zero-indexing scheme versus the Label values starting at one. This difference means that if `label_values[2]` equals one, the label is `BUILDING(3)`.

Second, the image's annotation uses twelve 32-bit floating point values (48 bytes). You can find the 1,234[th] image's annotation by seeking to offset `1234 * 48` and reading the next 48 bytes. The easiest way to parse those bytes is with the built-in `unpack` function in the `struct` module:

```
import struct
BYTES_IN_32BIT = 4
LABEL_COUNT = 12


def read_annotations(dataset):
    annotations = []
    index = 0
    with open('./raw/%s/annotation.bin' % dataset,'rb') as f:
        data = f.read()
        for offset in range(0,len(data), LABEL_COUNT * BYTES_
IN_32BIT):
            chunk = data[offset:offset+LABEL_COUNT * SIZE_UINT]
            label_values = struct.unpack('f'*LABEL_COUNT,chunk)
```

```
                annotation = Annotation(index,label_values)
                annotations.append(annotation)
                index += 1
        return annotations
```

Declaring the Annotation class

The Annotation class bundles together the information about a specific image. You'll initialize the object with the current index and label_values:

```
class Annotation:
    def __init__(self, index, label_values) -> None:
        self.__index= index
        self.__values = label_values

    @property
    def label(self):
        return self.__get_label()

    @property
    def index(self):
        return self.__index

    @property
    def values(self):
        return self.__values
```

Next, we'll add a property that maps the image's index to the local filename. Each file is the six-digit padded index within a folder named after the four-digit prefix. For example, the 12,345th image is at the ./0012/012345.jpg relative path:

```
    @property
  def file_name(self):
 file= str(self.__index).zfill(6)
        dir = str(int(self.index / 1000)).zfill(4)
        return os.path.join(dir,file+'.jpg')
```

Finally, we'll define the __get_label function to map the one-hot encoding to the Label enumeration class:

```
def __get_label(self):
    value = 1
    for x in self.__values:
        if not x == 1:
            value += 1
            continue
        else:
            break
    return Label(value)
```

Validate parsing the file

The final step is to verify that we've read the annotation.bin files correctly. This step requires counting the total labels and comparing them against the dataset's home page:

```
def confirm_annotations(annotations):
    counts = {}
    for ann in annotations:
        if not ann.label.value in counts:
            counts[ann.label.value] =1
        else:
            counts[ann.label.value] +=1

    keys = sorted(list(counts.keys()))
    for key in keys:
        print('[%s] %d' % (Label(key),  counts[key]))
```

Run the confirm_annotations function for train_annotations:

```
print('Training Set\n%s' % ('='*12))
confirm_annotations(train_annotations)
```

This command will output the following:

```
Training Set
============
[Label.PERSON] 4856
```

```
[Label.CAR] 3830
[Label.BUILDING] 2085
[Label.WINDOW] 4098
[Label.TREE] 1846
[Label.SIGN] 951
[Label.DOOR] 830
[Label.BOOKSHELF] 391
[Label.CHAIR] 385
[Label.TABLE] 192
[Label.KEYBOARD] 324
[Label.HEAD] 212
[Label.CLUTTER] 20000
```

Then, invoke `confirm_annotations` for `test_annotations`:

```
print('Test Set\n%s' % ('='*12))
confirm_annotations(test_annotations)
```

Running this snippet will return the following output:

```
Test Set
============
[Label.PERSON] 1180
[Label.CAR] 974
[Label.BUILDING] 531
[Label.WINDOW] 1028
[Label.TREE] 494
[Label.SIGN] 249
[Label.DOOR] 178
[Label.BOOKSHELF] 100
[Label.CHAIR] 88
[Label.TABLE] 54
[Label.KEYBOARD] 75
[Label.HEAD] 49
[Label.CLUTTER] 5000
```

Finally, compare these counts against the official expected values:

#	Object class	Instances in training set	Instances in test set
1	Person	4,885	1,180
2	Car	3,829	974
3	Building	2,085	531
4	Window	4,097	1,028
5	Tree	1,846	494
6	Sign	954	249
7	Door	830	178
8	Bookshelf	391	100
9	Chair	385	88
10	Table	192	54
11	Keyboard	324	75
12	Head	212	49
13	clutter	20,000	5,000
	total number of images	40,000	10,000

Table 10.5 – Expected label counts

Restructure the files

It's common for ML practitioners to organize their datasets into train/test folders with one child folder per label (e.g., `/data/train/label/file_name.jpg`). This pattern appears so frequently that the mainstream frameworks even offer utilities for initializing labeled datasets. For instance, they would understand that `a_image_1.jpg` is a `person`.

TensorFlow's `image_dataset_from_directory` and PyTorch's `ImageFolder` are examples of this behavior by efficiently batching the files, normalizing their size, and assigning labels using the folder names:

```
train/
...person/
......a_image_1.jpg
......a_image_2.jpg
...car/
......b_image_1.jpg
......b_image_2.jpg
```

The LabelMe-12 dataset doesn't follow this behavior, which adds undifferentiated heavy lifting to any data loading. Let's take a look at a simple and elegant approach to remapping the files with minimal I/O.

Linux-based operating systems support a feature called hard linking, which lets us access the same physical file from two different names. You can use this capability to avoid moving or duplicating the images into the expected folder structure.

Let's create two new folders, ./data/training and ./data/testing, with one child folder for every Label value:

```
from os import symlink, path, mkdir, link, remove
MAX_LABEL_VALUE=13
if not path.exists('./data'):
    mkdir('./data')
for ds, annotations in [('train', train_annotations),('test',
test_annotations)]:
    in_dir = './raw/%s' % ds
    out_dir = './data/%s' % ds
    if not path.exists(out_dir):
        mkdir(out_dir)

    for ix in range(0,MAX_LABEL_VALUE+1):
        label = Label(ix).name
        subdir = path.join(out_dir, label)
        if not path.exists(subdir):
            mkdir(subdir)
```

Next, iterate through train_annotations and test_annotations to create the hard links:

```
    for ann in annotations:
        src = path.join(in_dir,ann.file_name)
        dst = path.join(out_dir, ann.label.name, path.
basename(ann.file_name))
        if path.exists(dst):
            os.remove(dst)
        link(src,dst)
```

You can verify the behavior by randomly sampling a few files:

```
from IPython.display import Image
Image(filename='./data/train/BUILDING/000002.jpg')
```

Figure 10.15 – Building 000002.jpg

Load the dataset

Import the folder using the following command:

```
import tensorflow as tf
training_images = tf.keras.preprocessing.image_dataset_from_
directory(
    './data/train/',
    labels='inferred',
    label_mode='int',
    color_mode='grayscale',
    image_size=(256,256),
    follow_links=True,
    seed=0)

testing_images = tf.keras.preprocessing.image_dataset_from_
directory(
    './data/test/',
    labels='inferred',
    label_mode='int',
    color_mode='grayscale',
    image_size=(256,256),
```

```
        follow_links=True,
        seed=0)
```

You will see the following output:

```
Found 40000 files belonging to 14 classes.
Found 10000 files belonging to 14 classes.
```

Finally, you can render those images and begin exploring:

```
import matplotlib.pyplot as plt
plt.figure(figsize=(10, 10))
for images, labels in train_ds.take(1):
    print(len(images))
    for i in range(9):
        ax = plt.subplot(3, 3, i + 1)
        plt.imshow(images[i].numpy().astype("uint8"))
        plt.title(int(labels[i]))
        plt.axis("off")
```

Figure 10.16 – Bulk-loaded images

Quick recap

In this section, you learned how to parse the one-hot encoded `annotation.bin` binary file into the `Annotation` objects. You'll find many large datasets require you to unpack the metadata for size and performance purposes.

Summary

Amazon SageMaker is a comprehensive suite of services and capabilities that aims to bring ML to every developer, business analyst, and data scientist. You can mix and match its tooling with your existing environment or leverage it entirely for your ML model needs.

You also recreated Jian Yang's famous hotdog classifier using the AWS console and the built-in image classifier. While this step was more involved than *Chapter 4*, it didn't require a Ph.D. in data science. Hopefully, that gives you the confidence to experiment with twenty-four built-in algorithms plus the thousands available through AWS Marketplace.

You also observed how Amazon SageMaker's CloudWatch integration standardizes where you look for performance and troubleshooting information. Within a matter of clicks, you can drill down to specific error messages. Finally, the AWS CLI simplifies exporting the job definitions for programmatic access. Then, you learned that even datasets that use custom binary annotations wouldn't stop you from importing them into SageMaker and continuing to experiment further.

In the next chapter, you'll learn how to create hunab-in-the-loop workflows using Amazon A2I. Traditionally, human workforces needed to review *every* piece of content regardless of its quality. Amazon A2I reduces this waste by only involving your workforce when specific criteria and confidence thresholds are detected.

Part 5: Best Practices for Production-Ready CV Workloads

This fifth part consists of three cumulative chapters that will cover how to improve the accuracy of CV workloads using human reviewers, best practices to consider for your end-to-end CV pipelines, and the importance of establishing AI governance.

By the end of this part, you will understand how to use **Amazon Augmented AI (Amazon A2I)** to improve the accuracy of CV workflows, best practices for implementing cost optimization and security, and steps for applying AI governance.

This part comprises the following chapters:

- *Chapter 11, Integrating Human-in-the-Loop with Amazon Augmented AI*
- *Chapter 12, Best Practices for Designing an End-to-End CV Pipeline*
- *Chapter 13, Applying AI Governance in CV*

11

Integrating Human-in-the-Loop with Amazon Augmented AI (A2I)

Sometimes, the **machine learning** (**ML**) predictions are not accurate enough for your use case. Alternatively, you might require predictions with very high confidence (99% or higher) for use cases involving high-risk, high-impact decisions such as approving a loan application or taking down content on social media applications. In such cases, you will want to run the ML predictions through human reviewers. This is where **Amazon Augmented AI** (**Amazon A2I**) comes in. You can use Amazon A2I to easily build workflows that require human review for ML predictions. In other words, you can use Amazon A2I to set up human-in-the-loop workflows. Amazon A2I removes the complexity, cost, and heavy lifting involved in building human review workflows or managing a large group of human reviewers.

This chapter covers the following topics:

- Introducing Amazon A2I
- Learning how to build a human review workflow
- Leveraging A2I's built-in integration with Amazon Rekognition to review unsafe images

Technical requirements

For the examples in this chapter, you will require the following:

- Access to an active AWS account with permissions to access Amazon SageMaker and Amazon Lookout for Vision
- PyCharm or any Python IDE

- All the code examples for this chapter can be found on GitHub at `https://github.com/PacktPublishing/Computer-Vision-on-AWS`

A Jupyter notebook is available to run the example code from this chapter. You can access the most recent code from this book's GitHub repository, `https://github.com/PacktPublishing/Computer-Vision-on-AWS`. Clone this repository to your local machine using the following command:

```
$ git clone https://github.com/PacktPublishing/Computer-Vision-
on-AWS
$ cd Computer-Vision-on-AWS/11_AmazonA2I
```

Additionally, you will need an AWS account and Jupyter Notebook. *Chapter 1* contains detailed instructions for configuring the developer environment.

Introducing Amazon A2I

Building a human review system is quite complex, time-consuming, and expensive. Typically, it requires you to build custom applications to implement workflows and manage review tasks and consolidate results. Additionally, it needs to handle the human reviewer, who will work on the assigned tasks and submit results. Amazon A2I streamlines this process. It provides built-in human review workflows for common use cases, or you can build your own workflow.

With Amazon A2I, you can build the following workflow. Your input data will be sent to an AWS AI service (such as Textract or Rekognition) or custom ML models (hosted with SageMaker or self-managed endpoints). You can set up rules to send high-confidence predictions to client applications immediately and send low-confidence results to human reviewers via a human review workflow setup with Amazon A2I. Once the human reviewers review the predictions, their responses will be consolidated and stored on S3. The client application can then utilize the consolidated results from S3 to take further actions:

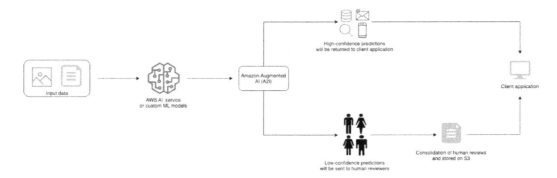

Figure 11.1: Human-in-the-loop with Amazon A2I

Amazon A2I has native integration with Amazon Textract and Amazon Rekognition for certain task types, where the AWS service will create a human loop on your behalf when the defined conditions (such as the confidence threshold) are met. If you're using custom ML models to generate predictions, you can start a human loop using the `StartHumanLoop` API of Amazon A2I. It will kick off the workflow and send the task to human reviewers.

Core concepts of Amazon A2I

There are three core concepts of Amazon A2I that you will need to know before incorporating Amazon A2I with your applications:

- **Task types** – The task type defines the workflow and how Amazon A2I will be integrated. There are two built-in task types: Amazon Textract key-value pair extraction and Amazon Rekognition image moderation. You can also create custom task types to integrate human review in any AI/ML workflow.

- **Human review workflow or flow definition** – The human review workflow defines three things: 1) the human reviewer team – whom you will send the review tasks, 2) the worker task template – which defines the worker UI or task UI that provides interactive tools for workers to use for completing tasks, and 3) instructions about how human reviewers should complete the review tasks.

- **Human loops** – A human loop is a single review job that will be worked upon by workers. The human loop kicks off the human review workflow and sends the input data to human reviewers as assigned tasks. You can specify how many workers will be sent a task to review a single data object. For example, you can specify that three workers will be sent the same image for an image classification job. Having multiple workers per data object helps improve label accuracy.

For built-in task types, you provide activation conditions under which the human loop should be initiated. For example, Amazon Textract can identify relationships between detected text items. You can use the human review workflow to send a document to a worker to verify key-value pair extraction when Amazon Textract's confidence is too low for specific form keys. For a custom task type, the human loop is created when you call `StartHumanLoop`.

Now that we know the core concepts of Amazon A2I, in the next two sections, we will leverage them to build a human review workflow.

Learning how to build a human review workflow

Before you get started with creating a human review workflow with Amazon A2I, you will need an S3 bucket in the same AWS Region to collect output data for the workflow.

Creating a labeling workforce

To build a human review workflow, you will need to define the labeling workforce who will label your dataset. As we learned in *Chapter 9*, you can choose either a public (Mechanical Turk), private (in-house), or vendor (AWS Marketplace/third-party) workforce to review your dataset. You can create and manage workforces from Amazon Ground Truth's console. If you have an existing workforce in Ground Truth, you can use the same workforce to review predictions.

Setting up an A2I human review workflow or flow definition

As we learned in the *Core concepts of Amazon A2I* section, next, we'll create a human review workflow to define the task types, the activation conditions if you're using a built-in task type, task template, and use the labeling workforce from the last step. You can use default templates for built-in task types, or you can design your own custom template. For custom templates, you can get started with the sample task UIs available on GitHub (`https://github.com/aws-samples/amazon-a2i-sample-task-uis`).

Initiating a human loop

Once the human review workflow has been created, you're ready to assign tasks to workers. You will be using API operations to start a human loop. If you're using built-in task types, you will use the AWS AI service API, such as `AnalyzeDocument` with Amazon Textract or `DetectModerationLabels` with Amazon Rekognition, and pass `HumanLoopConfig` as a parameter. Amazon Textract or Amazon Rekognition will initiate a human loop when the activation conditions, defined in a human review workflow, are met.

Now that we know what's required to build a human review workflow with Amazon A2I, let's create a human review workflow integrated with Amazon Rekognition to review unsafe images.

Leveraging Amazon A2I with Amazon Rekognition to review images

In this section, we will learn how to use Amazon A2I with Amazon Rekognition to review images. We will use multiple images containing moderation labels to see Amazon A2I in action.

Step 1 – Collecting your images

Upload the sample images from the book's GitHub repository. You can complete this step using the following command:

```
$ aws s3 sync 11_AmazonA2I/images s3://cv-on-aws-book-xxxx/
chapter_11/images --region us-east-2
```

> **Important note**
> To collect the sample images, you can use the same S3 bucket you created in *Chapter 2*.

Now, navigate to Amazon SageMaker on the AWS Management Console at `https://us-east-2.console.aws.amazon.com/a2i/home?region=us-east-2#`.

Step 2 – Creating a work team

Expand **Ground Truth** in the left-hand navigation pane, and select on **Labeling workforces**:

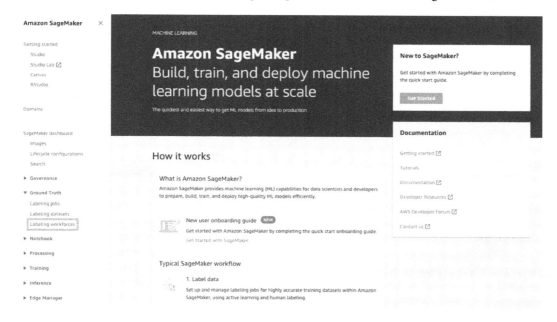

Figure 11.2: The Amazon SageMaker console

For this example, we will be using a private workforce. If you created a workforce for labeling images in *Chapter 9*, you can use the same workforce group. Otherwise, you can create a new workforce by following the instructions in *Chapter 9*:

Figure 11.3: Worker team configuration

In this example, we will use the existing worker team to review input data.

Step 3 – Creating a human review workflow

Next, we need to create a human review workflow:

1. Select on **Human review workflows** under the **Augmented AI** section of the SageMaker console. Select on **Create human review workflow**:

Figure 11.4: Creating a human review workflow in the SageMaker console

2. Under the **Workflow settings** section, you will need to provide a name for the workflow, the S3 bucket to store the output of the human review, and the IAM role to grant A2I permissions to call other services on your behalf:

Figure 11.5: Creating a human review workflow

3. If you don't have an existing role, select **Create a new role**. You can limit the IAM role permission to specific S3 buckets or allow access to any buckets. In this example, you can just provide the name of the S3 bucket you are using:

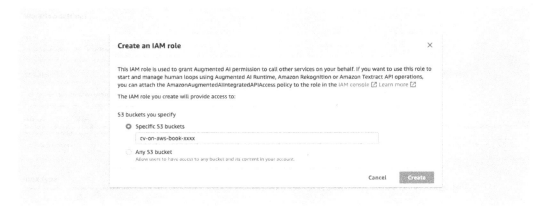

Figure 11.6: Creating an IAM role for A2I

The newly created IAM role only has permission to access the S3 bucket and objects within the bucket. Once the IAM role has been created, you will see an info wizard suggesting that you attach a managed IAM policy, named `AmazonAugmentedAIIntegratedAPIAccess`, to the role.

4. Navigate to the IAM console, and select on **Roles** in the navigation pane (`https://us-east-1.console.aws.amazon.com/iamv2/home#/home`):

Figure 11.7: Attaching permissions to the IAM role

5. In the **Permissions** tab, select **Add permissions** and then select **Attach policies**. This will open all the permission policies in your AWS account.

6. Search for `AmazonAugmentedAIIntegratedAPIAccess`, select on the checkbox to select the policy, and then select on **Attach policies**. This should give you confirmation that the policy was successfully attached to the role. Navigate back to the Amazon A2I configuration.

7. Next, we need to configure the task type. This provides options to select built-in task types with Amazon Textract and Amazon Rekognition along with custom task types. We will select **Rekognition – Image moderation**:

Figure 11.8: The task type configuration

8. As you selected a built-in task type, you need to configure the conditions for invoking the human review. You can either trigger the workflow based on the label confidence score by providing a threshold, or you can randomly send a sample of images and their labels to a human for review.

For this example, you can select the **Trigger human review for labels identified by Amazon Rekognition based on the label confidence score** condition and set the threshold between 0 and 95. This configuration means that A2I will invoke a human loop for any labels identified by Amazon Rekognition with a confidence score in the range of 0 to 95. Essentially, you're only invoking a human review if the confidence score for identified labels is below 95:

Amazon Rekognition-Image moderation - Conditions for invoking human review

Learn more about using Amazon Augmented AI with Amazon Rekognition [↗]

☑ Trigger human review for labels identified by Amazon Rekognition based on the label confidence score.
Labels will be sent for human review.

Threshold
Trigger a human review for any labels identified with a confidence score in the following range:

between [0] and [95]

Minimum value is 0. Maximum value is 100.

☐ Randomly send a sample of images and their labels to humans for review.
For each image sent, all labels identified by Amazon Rekognition for that image will be sent for human review.

Figure 11.9: Configuring the conditions for invoking a human review

9. Next, we need to provide task template information. We can either create this from a default template, or you can use your own template:

Worker task template creation
Create a template or use a template created in the Worker task templates section of the console. Learn more about worker task templates [↗]

Template type

◉ Create from a default template	○ Use your own template

Template name

default-image-moderation-template

The name must be lowercase, unique within the Region in your account, and can have up to 63 characters. Valid characters: a-z, 0-9, and - (hyphen)

Figure 11.10: Worker task template configuration

10. Next, we will confirm a template design and provide a task description for your workers. The default template comes with example instructions, so you can update them as needed. You can select on **See a sample worker task** to test how tasks will be presented to workers:

Figure 11.11: Worker task template configuration

11. At last, you will select worker types to perform the review. In this example, you will select the **Private** worker type and select the existing team or the new team you created. In the **Additional configuration** section, you can configure settings such as workers per object, task timeouts, and task expiration time:

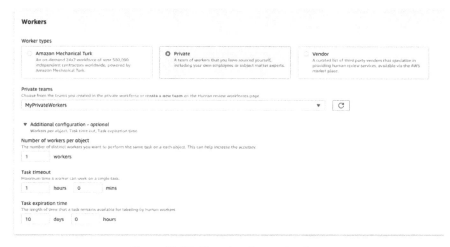

Figure 11.12: Choosing the worker type

12. Next, select on **Create**, and it should create a human review workflow with the **Active** status. You can copy and save the Workflow ARN, as you will need it for the next step.

Step 4 – Starting a human loop

As you complete the creation of the human review workflow, next, you will test the human-in-the-loop setup. To start a human loop, you will need to use the API operation. For this example, we will use Rekognition's `DetectModerationLabels` API.

The following request example uses the AWS CLI to make the API call:

```
aws rekognition detect-moderation-labels \
    --region us-east-2 \
    --image "S3Object={Bucket='cv-on-aws-book-xxxx',
Name='chapter_11/images/image1.jpg'}" \
    --human-loop-config '{"HumanLoopName": "image-review-
human-loop", "FlowDefinitionArn": "arn:aws:sagemaker:us-
east-2:xxxxxxxxxxxx:flow-definition/unsafe-image-review-
workflow", "DataAttributes": {"ContentClassifiers":
["FreeOfPersonallyIdentifiableInformation",
"FreeOfAdultContent"]}}'
```

In the preceding request, we are providing the region, the input image location, and the human loop config. We can provide a name to the human loop, and refer to the human review workflow ARN (or flow definition ARN) from *step 3*. Lastly, you can optionally provide content classifiers to declare that the images are free of PII information and adult content; however, they are not needed in your example with a private workforce. These attributes are needed if you are using Amazon Mechanical Turk for workers.

The response of the preceding CLI command will look like the following:

```
{
    "ModerationLabels": [
        {
            "Confidence": 97.1685562133789,
            "Name": "Alcoholic Beverages",
            "ParentName": "Alcohol"
        },
        {
            "Confidence": 97.1685562133789,
            "Name": "Alcohol",
            "ParentName": ""
```

```
        }
    ],
    "ModerationModelVersion": "6.0",
    "HumanLoopActivationOutput": {
        "HumanLoopActivationReasons": [],
        "HumanLoopActivationConditionsEvaluationResults":
"{\"Conditions\":[{\"And\":[{\"ConditionType\":\"Moderation-
LabelConfidenceCheck\",\"ConditionParameters\":{\"Modera-
tionLabelName\":\"*\",\"ConfidenceLessThan\":95.0},\"Evalua-
tionResult\":false},{\"ConditionType\":\"ModerationLabelCon-
fidenceCheck\",\"ConditionParameters\":{\"ModerationLabel-
Name\":\"*\",\"ConfidenceGreaterThan\":0.0},\"EvaluationRe-
sult\":true}],\"EvaluationResult\":false}]}"
    }
}
```

Looking at the preceding response, it looks like Amazon Rekognition detected the presence of **alcoholic beverage** labels with over 97% confidence. It also showed that human loop activation conditions were not met (as we defined to only send images to workers if the confidence score was below 95). It needs to meet both the defined conditions: the confidence score being greater than 0 and the confidence score being less than 95. Evaluation for one condition was true and the other was false, so the final result is false and the human loop wasn't triggered.

Let's try with another sample image and make the API call:

```
aws rekognition detect-moderation-labels \
    --region us-east-2 \
    --image "S3Object={Bucket='cv-on-aws-book-xxxx',
Name='chapter_11/images/image2.jpg'}" \
    --human-loop-config '{"HumanLoopName": "image-review-
human-loop", "FlowDefinitionArn": "arn:aws:sagemaker:us-
east-2:xxxxxxxxxxxx:flow-definition/unsafe-image-review-
workflow", "DataAttributes": {"ContentClassifiers":
["FreeOfPersonallyIdentifiableInformation",
"FreeOfAdultContent"]}}'
```

You should get the following response. As you can see in the preceding response, since the confidence score for the label is about 90%, the human loop was invoked:

```
{
    "ModerationLabels": [
        {
```

```
                "Confidence": 90.10928344726562,
                "Name": "Alcoholic Beverages",
                "ParentName": "Alcohol"
            },
            {
                "Confidence": 90.10928344726562,
                "Name": "Alcohol",
                "ParentName": ""
            }
        ],
        "ModerationModelVersion": "6.0",
        "HumanLoopActivationOutput": {
            "HumanLoopArn": "arn:aws:sagemaker:us-east-2:
xxxxxxxxxxxx:human-loop/image-review-human-loop",
            "HumanLoopActivationReasons": [
                "ConditionsEvaluation"
            ],
            "HumanLoopActivationConditionsEvaluationResults":
"{\"Conditions\":[{\"And\":[{\"ConditionType\":\"Moderation-
LabelConfidenceCheck\",\"ConditionParameters\":{\"Modera-
tionLabelName\":\"*\",\"ConfidenceLessThan\":95.0},\"Evalu-
ationResult\":true},{\"ConditionType\":\"ModerationLabelCon-
fidenceCheck\",\"ConditionParameters\":{\"ModerationLabel-
Name\":\"*\",\"ConfidenceGreaterThan\":0.0},\"EvaluationRe-
sult\":true}],\"EvaluationResult\":true}]}"
        }
    }
}
```

Let's head to the labeling portal. You can find sign-in URL information in the Amazon SageMaker console on the **Labeling workforces** page under **Ground Truth**. Once you sign in, you will notice the pending job in the portal:

Figure 11.13: The worker portal

You can select on **Start working**. You should see a screen similar to the following:

Figure 11.14: Worker task (1)

It presents the input data (the photograph) along with the labels Rekognition identified and the instructions for workers to follow:

Figure 11.15: Worker task (2)

As the photograph represents alcoholic beverages and alcohol, you can select those categories and select on the **Submit** button.

Step 5 – Checking the human loop status

You can call the Amazon Rekognition API for all the sample images provided in the GitHub repo. Keep in mind, you will need to provide a unique name for each human loop. Once you have completed analyzing all the images, you can go to the Amazon A2I portal on the SageMaker console and check out all the instances of when the human loop was invoked:

Figure 11.16: Invoked human loop summary on the Amazon SageMaker console

You can select on any of the human loops to get more details, such as the status and output location:

Figure 11.17: The human loop details

Select on the output location to capture the S3 URI for the object to analyze output data in the next section.

Step 6 – Reviewing the output data

Once the review for the entire dataset has been completed by your workforce, you can direct your client applications to review and process the output data. To review the output data locally, you can just download the output file from the S3 bucket location you specified in the human review workflow. You can use the following S3 API to copy the object locally on your machine or into Jupyter Notebook:

```
aws s3 cp s3://cv-on-aws-book-xxxx/chapter_11/review_data/
unsafe-image-review-workflow/2022/12/22/19/25/33/image-review-
human-loop/output.json .
```

You can open the downloaded output.json file to review the worker response. The following output has been truncated:

```
{
"awsManagedHumanLoopRequestSource":"AWS/Rekognition/
DetectModerationLabels/Image/V3",
    "flowDefinitionArn":"arn:aws:sagemaker:us-east-
2:xxxxxxxxxxxx:flow-definition/unsafe-image-review-workflow",
    "humanAnswers":[
        {
            "acceptanceTime":"2022-12-22T19:48:48.555Z",
            "answerContent":{
                "AWS/Rekognition/DetectModerationLabels/Image/V3":{
                    "moderationLabels":[
                        {
                            "name":"Alcoholic Beverages",
                            "parentName":"Alcohol"
                        },
                        {
                            "name":"Alcohol",
                            "parentName":""
                        }
                    ]
                }
            },
            "submissionTime":"2022-12-22T19:49:34.629Z",
            "timeSpentInSeconds":46.074,
            "workerId":"fd357806f5b5ea70",
            "workerMetadata":{
                "identityData":{
                    "identityProviderType":"Cognito",
                    "issuer":"https://cognito-idp.us-east-2.
amazonaws.com/us-east-2_Ldbmzk97S",
                    "sub":"5fc3aded-fdf8-4cc3-923b-94f7d2f98c36"
```

```
            }
        }
    }
],
. . .
. . .
. . .
```

Key information in the output data is located within the humanAnswers key. It shows at which time the worker accepted the task, their response to moderation labels, the time they spent on this task, and the worker ID. You can use the information located in the answerContent key accordingly in your application to process the input data. The output file also includes the input content, such as the image location and name, and a response from Amazon Rekognition with the confidence score for each label.

Summary

In this chapter, we covered what Amazon A2I is and how you can use it to build human-in-the-loop workflows. We discussed the core components of Amazon A2I and the process to build a human review workflow. In the end, we used a built-in human review workflow for content moderation and engaged human reviewers when Amazon Rekognition was unable to make a high-confidence prediction.

In the next chapter, we will focus on the design principles you should use while building **computer vision (CV)** applications that will help you reduce costs, increase security, and support a growing user base.

12
Best Practices for Designing an End-to-End CV Pipeline

In the previous chapter, we introduced **Amazon Augmented AI (A2I)**. We discussed the importance of a human-in-the-loop workflow and walked through a code example while using Amazon Rekognition to analyze unsafe images.

Throughout this book, we've covered several real-world CV use cases. We've also discussed AWS AI/ML services in detail, including Rekognition, Lookout for Vision, and SageMaker. To recap, AWS AI/ML services are managed services, which means that the undifferentiated heavy lifting of patching, upgrading, and maintaining servers and hardware is removed. AWS AI services are composed of serverless architecture that scales automatically. Rekognition includes pre-trained capabilities, which means that steps such as preparing and transforming data, selecting an algorithm to train the model, and tuning the model are not required. SageMaker has some distinguishing characteristics compared to AWS AI services. It is also a managed service, but ML development experience is required. The data preparation, model selection and training, tuning, and deployment and monitoring steps are all required. However, SageMaker is composed of many comprehensive capabilities to address each step of the ML workflow.

In this chapter, we will provide a detailed overview of the best practices to consider when designing an end-to-end CV pipeline. For each step of the ML life cycle, we will define guidelines and processes for architecting a CV system.

In this chapter, we will discuss the following:

- Defining a problem that CV can solve and processing data
- Developing a CV model
- Deploying and monitoring a CV model
- Developing an MLOps strategy
- Using the AWS Well-Architected Framework

Defining a problem that CV can solve and processing data

The critical first step before designing a CV system is to define the problem you're trying to solve and the desired business outcomes. Involve the relevant stakeholders to identify any pain points and a solution that could be solved with CV. Understand the constraints and requirements to solve the problem and evaluate the available data. Specify what success looks like and identify the **key performance indicators (KPIs)**. Consider the costs, resources available, and security and compliance requirements. Here is a list of additional questions to ask during this process:

- What are the goals of the solution? Is it to increase revenue, provide actionable insights, reduce manual processes, or something else?
- What is the timeline for the solution?
- How will the model be integrated into downstream systems?
- What type of data will be processed?
- What are the data privacy requirements and how will these be addressed?
- What resources are required – networking, compute, storage?
- How will model training be conducted and how will it be tested?
- What performance metrics need to be optimized?
- How will the model be monitored for bias and fairness?

Once you've identified a problem that CV can solve and outlined your desired business goals, you move to the data processing stage. Sampling is a key step for gathering data to train a CV model. To help determine the best method for data sampling, you need to define your business objectives and what type of data is required. Other considerations include analyzing the population and sample size available, as well as understanding the cost, time, and resource constraints required for collecting the data. Also, the process should account for any regulation requirements, compliance, and ethical considerations. Several sampling methods can be used to help prevent biases and improve the efficiency of the training data available. Here is an overview of several methods:

- **Simple random sampling**: This is the simplest method of probabilistic sampling. In theory, each sample has an equal chance of being randomly selected. However, if a rare class is only representative of a small percentage of the data, then that class may not be selected.
- **Systematic sampling**: A probabilistic sampling method where data is selected at the same interval, such as every k^{th} element being selected.
- **Cluster sampling**: This method divides a population into groups, called clusters, usually based on geographic location. Each of these clusters is then randomly selected.
- **Stratified sampling**: This is similar to cluster sampling, but the population is put into groups or strata based on a certain characteristic.

- **Convenience sampling**: A nonprobabilistic (non-random) sampling method where data is selected based on what is available.

- **Weighted sampling**: Each item in a sample is given a weight and the probability of it being selected is determined by its weight.

- **Reservoir sampling**: An algorithm for dealing with streaming data, where every item in the stream needs to have an equal chance of being randomly selected.

Data labeling or data annotation is an important process for preparing datasets for supervised learning. To better understand how to improve label quality, review *Chapter 9*, where we provided an example using Amazon SageMaker Ground Truth.

After you have collected and labeled your data, it is important to split your data into training, validation, and test datasets. This ensures that the model is generalizable, which means that it can be applied to future, unseen data. A training dataset is meant for learning the model's parameters. The validation dataset helps to tune the hyperparameters, while the test dataset gives you a reliable estimation of the model's performance on the unseen data. There is no right answer to how much data should be allocated for training, validation, and test datasets. There should be a right balance and this balance can vary for every problem. If your test dataset is too small, your performance statistic will have high variance and you will have an unreliable estimation of model performance. If your training dataset is too small, your model parameters will have high variance. Generally, a good approach is a 60/20/20 training, validation, and test split.

In *Chapter 1*, we provided an overview of a few data preprocessing techniques for CV, including image resizing, image masking, and grayscaling. Another useful technique is normalization, where features are centered and rescaled, which results in a similar data distribution. Data augmentation is a process that helps prevent overfitting by increasing the size of the training data available, without requiring additional data collection. Transformations can be applied to images such as rotation, flipping, cropping, adding contrast, and scaling.

The purpose of feature engineering is to help improve the performance of CV models; it improves model accuracy by extracting the most important features and makes the model more interpretable, as well as more robust to data changes. Techniques such as feature extraction reduce dimensionality, which reduces the number of features, making it easier for a model to learn. When you use a deep learning algorithm, deep neural networks automatically extract features from an image. If you are not using deep learning for feature extraction, the following are a few feature engineering techniques that are useful in CV:

- **Image filtering**: This applies various filters to an image, such as changing colors, blurring, and enhancing to remove noise.

- **Image segmentation**: This involves dividing an image into different regions or segments so that the pixels in the segment are representative of similar characteristics.

- **Scale-Invariant Feature Transform** (**SIFT**): This is an algorithm that locates the features in an image.

- **Speeded-Up Robust Features** (**SURF**): This is similar to SIFT, but it is more performant.

- **Histograms of Oriented Gradients** (**HOG**): This describes the shape or structure of an object. It is also used in image recognition.

- **Local Binary Patterns** (**LBP**): This is used to describe the texture of an object in classification and object detection use cases.

- **Image moment invariants**: This describes the shape, size, and orientation of objects. It is useful in identity analysis.

- **Image pyramids**: These reduce the resolution or resize an image. Pyramids are a collection of images stemming from the original image.

In this section, we summarized the importance of identifying a CV problem to be solved and the desired business goals. We provided a list of questions to consider to assist during this process. We also covered sampling methods, the importance of splitting the data into training, validation, and test datasets, provided an overview of data processing techniques, and described feature engineering techniques for CV. In the next section, we will detail the best practices for training, evaluating, and tuning a CV model.

Developing a CV model

Once you have a training dataset, you move to the model development stage, which includes training, evaluating, and tuning a model. Algorithm selection is not a one-size-fits-all process and model development is iterative. In this section, we will not dive deep into training algorithms, but will instead highlight a few best practices and the training options available with AWS AI/ML services.

Training

In *Chapter 1*, we briefly covered a few types of algorithms that are often used to solve CV use cases, such as object detection, classification, and segmentation. There are many algorithms available to solve CV problems and choosing the best one depends on a few different factors, including the type of problem you are trying to solve, the data you have available, and your specific performance requirements. Deep learning algorithms often work well for CV, but other more traditional ML algorithms such as **Support Vector Machines** (**SVMs**), random forests, and **k-nearest neighbors** (**k-NN**) are sometimes useful for solving classification problems. Here is a summary of some of the deep learning algorithms available for CV:

- **CNN**: This is commonly applied to image recognition and classification tasks and extracts features from images.

- **Recurrent Neural Network (RNN)**: This is useful for image captioning and video classification. It considers the temporal relationships between DataFrames.

- **Generative Adversarial Network (GAN)**: This is often used for image synthesis and style transfer. It is a generative model that generates new output images from the provided input.

- **R-CNN**: This algorithm detects objects in an image using region proposal networks. Usually, they are quite accurate but computationally expensive.

- **YOLO**: This algorithm is used for real-time object detection. It uses a single neural network for predicting the class and location of objects in an image. It is often faster than R-CNN but less accurate.

- **SSD**: This is a single-shot detector that divides the image into a grid to detect objects. It is usually faster than R-CNN and more accurate than YOLO.

Transfer learning can be used for CV problems where a pre-trained deep neural network that was trained for a different task can be used as a starting point to complete a new task. You can fine-tune the pre-trained network for the new task, which reduces the amount of data, time, and resources that are required for training. For example, you could use a pre-trained image classification model that was trained on a large dataset and transfer that knowledge to your new model so that you do not have to start from scratch, gathering data and training a model. One of the benefits of Rekognition Custom Labels is that it uses transfer learning to automatically select the best algorithm for model training based on the data you provide and performs hyperparameter optimization. SageMaker provides built-in algorithms for CV. For image classification, object detection, and semantic segmentation tasks, there are options to fully train a model, use transfer learning, or use pre-trained models with SageMaker.

Evaluating

Performance metrics are used to evaluate a model. For classification models, one evaluation metric that is often used is accuracy. The formula for accuracy is the number of correct predictions divided by the total number of predictions. Accuracy is a suitable evaluation metric when the dataset is balanced. However, if a class imbalance occurs – you have a much higher proportion of some classes than other classes – measuring accuracy proves problematic. In the case of dealing with an imbalanced dataset, you could use the F1 score as a better evaluation metric. The F1 score is the harmonic mean between precision and recall. Precision is the ratio of true positives (the model correctly predicts the class) to all selected predictions. Recall, or sensitivity as it is defined in mathematics, is the ratio of selected true positives to all true positives. Amazon Rekognition Custom Labels provides each of these metrics for evaluation after testing your model. You can also view the confusion matrix of a classification model using the Rekognition Custom Labels SDK:

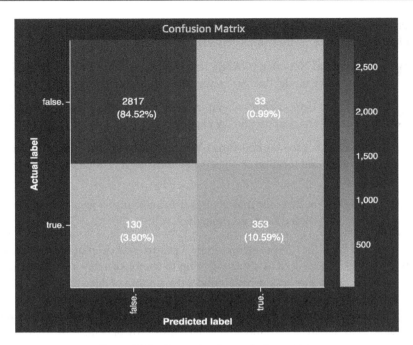

Figure 12.1 – Example of a confusion matrix

For objection detection models, evaluation metrics such as **average precision (AP)** and **mean average precision (mAP)** can be used. mAP and **mean average recall (mAR)** are metrics for image detection in Rekognition Custom Labels. Once the model has been evaluated, it can be fine-tuned and its hyperparameters are optimized to improve the model's performance.

Tuning

Hyperparameters describe structural information about a model and define properties such as learning rate and model complexity. They need to be decided before you can fit the model parameters. The following are several types of hyperparameter tuning methods:

- **Grid search**: Performs an exhaustive search on a specified set of hyperparameters that you define.

- **Random search**: Randomly searches for hyperparameters over a predefined range. It keeps searching until the desired accuracy is achieved.

- **Bayesian optimization**: Uses a Bayesian model to model the relationship between hyperparameters and the model's performance. The Bayesian model is updated after each evaluation and determines the hyperparameter values to test next.

- **Hyperband**: Selects the best combination of hyperparameters for a model based on a predefined metric. It randomly samples hyperparameters, trains models, and compares the performance of the models using the predefined metric.

These techniques are used in **Amazon SageMaker Automatic Model Tuning** (`https://aws.amazon.com/sagemaker/automatic-model-tuning/`) to find the best version of your model based on the hyperparameter ranges you set. The following diagram shows the steps for SageMaker Automatic Model Tuning:

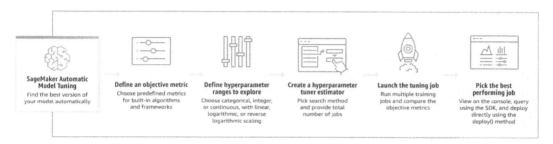

Figure 12.2 – SageMaker Automatic Model Tuning steps

Regularization is another tuning technique that prevents overfitting by artificially discouraging complex models. It penalizes complexity and often limits the flexibility of the model. Regularization techniques such as L1 (Lasso), which performs variable selection and parameter shrinkage, and L2 (Ridge), which only performs parameter shrinkage and regression assist with finding the optimal balance between a model underfitting and overfitting.

In this section, we discussed best practices for developing a CV model. We covered training algorithms, evaluation metrics, and hyperparameter tuning techniques. In the next section, we will discuss several testing strategies we can perform after deploying a CV model and also cover best practices for monitoring.

Deploying and monitoring a CV model

Once a CV model has been deployed, you should regularly evaluate its performance to establish when retraining is required and perform deployment testing before rolling out a new version. This ensures that models are delivering reliable results and that they are meeting your established business outcomes. Deployment testing helps you detect changes in a model's accuracy and identify any errors in a model's implementation before it is deployed to production.

Shadow testing

Shadow testing is a testing technique for evaluating the performance of a model before it's rolled out to production. A new (shadow) ML model is tested in a production environment without impacting actual user traffic. The shadow model's predictions are not used in the production application; instead, the shadow model runs alongside the existing production model. A copy of the inference requests is routed to the shadow model and its predictions are compared with the predictions of the production model. The advantage of shadow testing is that a shadow model's performance can be evaluated before rolling the model out to production. Amazon SageMaker supports shadow testing (`https://aws.amazon.com/sagemaker/shadow-testing/`) and automatically creates a live dashboard of metrics to assist with comparing the performance of the shadow and production models. This eliminates the need to build out your infrastructure for shadow testing, which is often a cumbersome task.

A/B testing

A/B testing is also a technique for comparing two different model versions. Unlike shadow testing, where a copy of the inference requests that were sent to the production models are also routed to the new model, the inference requests are split between sets of users. In other words, the new model version is only released to a subset of users and receives some of the inference requests. The performance of both model versions is then compared, before determining whether the new model should be deployed to production.

Blue/Green deployment strategy

A Blue/Green deployment strategy for ML includes two identical environments, with different model versions. The green environment contains the new model version, while the blue environment contains the existing model version. The green environment is deployed as a staging environment, while the blue environment remains in production. Traffic is shifted from the blue version to the green version gradually. If testing of the green version is successful, all traffic is then routed to the green version, and it becomes the production version. If issues are encountered with the green version, then traffic can be routed back to the blue version.

SageMaker provides deployment guardrails (`https://docs.aws.amazon.com/sagemaker/latest/dg/deployment-guardrails.html`) for Blue/Green deployments. This set of capabilities allows you to monitor performance metrics that you define and trigger alarms using Amazon CloudWatch. If an alarm is triggered, an auto rollback occurs, and traffic is shifted back to the blue model version. If an alarm is not triggered, then traffic stops being sent to the blue version and all traffic is routed to the green version. SageMaker manages the routing of traffic based on the traffic shifting mode that is defined: all-at-once, linear, or canary. The following diagram shows how traffic is shifted using SageMaker deployment guardrails:

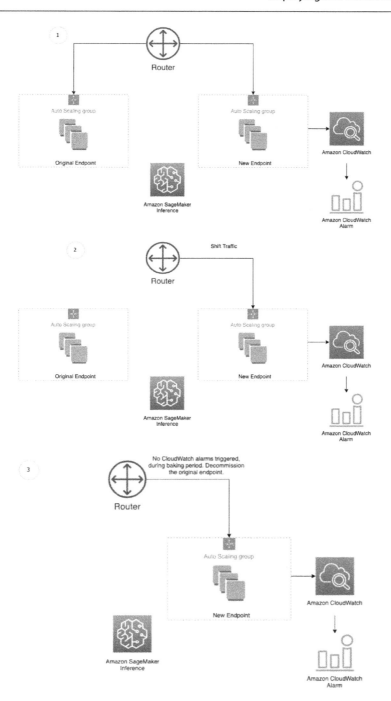

Figure 12.3 – Amazon SageMaker deployment guardrails steps

Monitoring

Once a CV model has been deployed, it is important to continuously enable monitoring to address any issues affecting its reliability, accuracy, and performance. Monitoring also helps with identifying and alerting on drift. Data drift occurs when the inference data deviates from the data used to train the model, decreasing model accuracy. In CV, there are a few methods that can be used for drift detection. Model-based methods use models to detect drift by comparing the performance of the models, while distance-based methods calculate the difference in distance between the distribution of the inference and training data.

Amazon SageMaker Model Monitor (`https://aws.amazon.com/sagemaker/model-monitor/`) can be used to monitor different types of drift, such as data quality, model quality, bias drift, and feature attribution drift. SageMaker Model Monitor integrates with **Amazon SageMaker Clarify** (`https://aws.amazon.com/sagemaker/clarify/`) to detect bias and promote model transparency and explainability. We will discuss SageMaker Clarify in more detail in the next chapter. SageMaker Model Monitor not only alerts on different types of drift but reports can be generated for visual analysis to assist with understanding the model's behavior. It also supports batch and real-time inference and has a built-in container for emitting metrics to Amazon CloudWatch.

In this section, we discussed strategies for testing and monitoring a CV model in production. We covered shadow testing, A/B testing, Blue/Green deployments, and the importance of continuously monitoring a model after deployment. We also detailed the capabilities available in SageMaker for deployment testing and model monitoring. In the next section, we will discuss how to develop an MLOps strategy.

Developing an MLOps strategy

Now that we've discussed the best practices to apply to each stage of the ML life cycle, how can we automate and streamline these processes? We can accomplish this by incorporating MLOps. What is MLOps? MLOps is related to DevOps in concept, where both practices focus on automating and accelerating applications or systems from development to production. The difference between the two is that the goal of DevOps is to deliver software applications, while the goal of MLOps is to deliver ML models. MLOps allows you to automate your ML workflows and create repeatable mechanisms to accelerate the processes for building, training, deploying, and managing ML models. You can leverage tools such as workflow automation software for orchestration and **continuous integration/ continuous delivery** (**CI/CD**) of your ML systems. Other components of MLOps include tracking lineage using a Model Registry. Also, monitoring models in production and providing corrective actions for retraining is another important step. Here is a diagram that highlights the MLOps steps and components on AWS when using SageMaker:

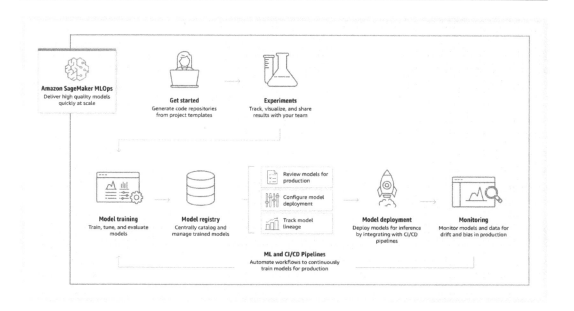

Figure 12.4 – MLOps with Amazon SageMaker

There are also steps you should consider when developing an effective MLOps strategy:

1. Determine and document your organization's business outcomes and the value that you want to derive from your MLOps strategy.

2. Take inventory of your current infrastructure requirements such as storage, compute, ML platform, resources, and tools for successful model development, deployment, and management.

3. Identify any pain points that you're experiencing with your current ML workflows.

4. Choose the tools and ML platforms you will use to build MLOps workflows.

5. Ensure that a process is implemented and metrics are defined for evaluating and monitoring models.

6. Establish an AI governance framework and address compliance and regulation requirements. We will define how to establish AI governance and its importance in much greater detail in the next chapter.

7. Understand that MLOps is not a one-time process, and that continuous feedback loops should be integrated within your teams and ML systems.

Now that we have defined the steps that should be considered for developing a comprehensive MLOps strategy, we will summarize the tools available on AWS for implementation.

SageMaker MLOps features

SageMaker Pipelines (`https://aws.amazon.com/sagemaker/pipelines/`) is a fully managed feature that allows you to automate and orchestrate steps of the ML workflow, including data processing, model training, and model evaluation. It also includes an SDK and a visual interface in SageMaker Studio. Each step of the pipeline can be executed automatically to accelerate model development and retraining. Once a model has been trained, it can be registered in SageMaker Model Registry (`https://docs.aws.amazon.com/sagemaker/latest/dg/model-registry.html`), which tracks the model versions and respective artifacts, including the lineage and metadata. It manages access to models and metadata to ensure that only correct permissions are given to authorized users. The Model Registry also manages the approval status of the model version for downstream deployment.

SageMaker Projects (`https://docs.aws.amazon.com/sagemaker/latest/dg/sagemaker-projects.html`) builds on Pipelines by implementing the model deployment steps and using the Model Registry, along with your existing CI/CD tooling, to automatically provision a CI/CD pipeline. Projects provides preconfigured MLOps templates that can also be customized to include the CI/CD tools that your organization prefers. It also helps with managing dependencies and code repositories, reproducing builds, and sharing artifacts across your organization. SageMaker Projects is provisioned using AWS Service Catalog (`https://aws.amazon.com/servicecatalog/`).

SageMaker Model Monitor is also integrated with Pipelines and Model Registry. This allows you to automate the process of maintaining the quality of models and acting quickly on issues that arise. SageMaker Experiments (`https://docs.aws.amazon.com/sagemaker/latest/dg/experiments.html`) is another feature that is integrated with Pipelines to automatically track and compare experiments. This is useful for comparing the performance of different configurations and models and reproducing a previous experiment.

MLOps is also a key component of ML governance. ML governance involves the processes, policies, and procedures for implementing responsible AI and ensuring the ethical use of data. MLOps helps enable the transparency, traceability, and auditability of your end-to-end pipeline. This promotes automation of the development and deployment of models in a secure and compliant manner.

Workflow automation tools

There are numerous workflow automation tools available that can be used to orchestrate and automate pipelines. The tools you choose depend on your requirements and the resources available within your organization. Here are a few options that can be integrated with AWS services:

- **Apache Airflow**: An open source platform to orchestrate tasks and automate pipeline workflows. **Amazon Managed Workflows for Apache Airflow** (**MWAA**) is one option.
- **Kubeflow**: This is a platform for composing, deploying, and managing ML workflows on Kubernetes.

- **AWS Step Functions**: This is a serverless option for building and automating ML workflows. It integrates with SageMaker.

In this section, we discussed the steps and tools to consider when implementing an MLOps strategy. We discussed fundamental MLOps best practices, but the tools and services we highlighted are merely suggestions to get you started. What tools you choose to use to build your ML pipelines will depend on a few factors, including your organizational requirements, the existing services you already have in use, and the skillsets you have available. We hope this section has helped you make informed decisions before architecting MLOps workflows. In the next section, we will dive deep into the AWS Well-Architected Framework from a CV perspective.

Using the AWS Well-Architected Framework

The AWS Well-Architected Framework (`https://docs.aws.amazon.com/wellarchitected/ latest/framework/welcome.html`) helps you design and evaluate your infrastructure to ensure it is secure, efficient, cost-optimized, reliable, and sustainable on AWS. The framework provides guidance and considerations for building and operating your workloads on AWS. It includes questions that assist you with identifying areas for improvement that focus on six pillars: cost optimization, operational excellence, reliability, performance efficiency, security, and sustainability. In addition, the AWS Well-Architected Machine Learning Lens (`https://docs.aws.amazon.com/ wellarchitected/latest/machine-learning-lens/machine-learning-lens. html`) is a resource for evaluating ML-specific workloads. Let's address the best practices to consider for each pillar when architecting CV workloads.

Cost optimization

Cost optimization includes understanding how to manage your resources, identifying usage, and minimizing costs. We mentioned earlier in this chapter that Rekognition and Lookout for Vision are serverless services. This means that there are no compute instances that you need to manage or the right size. For Rekognition Custom Labels and Lookout for Vision, you are charged for training and inference hours. To perform inference, you start your model and specify the number of compute resources (inference units) to provision. You are charged for the number of hours the model is running * the number of **inference units (IUs)**. It is important to stop the model when it is not in use since you are still charged, even if you are not making predictions. You can also auto-scale the IUs to account for changes in demand. Several other factors impact the number of IUs required, such as the complexity of the model; including higher-resolution images in your dataset will require more time for inference. Set up a testing environment to understand the throughput of your model and to calculate the IUs that need to be provisioned. Guidance on how to calculate the IUs required for Rekognition Custom Labels and Lookout for Vision is outlined in this blog: `https://aws.amazon.com/blogs/ machine-learning/calculate-inference-units-for-an-amazon-rekognition- custom-labels-model/`.

Here are some additional cost optimization best practices to consider:

- Implement a tagging strategy (`https://docs.aws.amazon.com/general/latest/gr/aws_tagging.html`) to monitor your usage and track your costs.

- Use AWS Budgets (`https://aws.amazon.com/aws-cost-management/aws-budgets/`) to define how much you want to spend or use a service and set up alerts if your cost or usage reaches a certain threshold.

- If you are using SageMaker, right-size your instances. Consider Amazon SageMaker Debugger (`https://aws.amazon.com/sagemaker/debugger/`) for profiling your training jobs and improving resource utilization. Use a life cycle configuration script to shut down idle notebook instances. Test pipelines locally before deploying on AWS using local mode (`https://docs.aws.amazon.com/sagemaker/latest/dg/pipelines-local-mode.html`).

Operational excellence

Operational excellence focuses on improving your processes and procedures to maximize business value. This includes defining business objectives within your organization and establishing processes for continuous improvement. Here are some design principles for operational excellence:

- Define performance metrics and KPIs and set up monitoring and alerting with CloudWatch.

- Implement data profiling and set up a process for improving data quality. You can get insights about your data with SageMaker Data Wrangler.

- Standardize consistency across development, testing, and production environments with **Infrastructure as Code (IaC)** using AWS CloudFormation.

Reliability

Reliability involves how well your workload can recover from failure and adapt to change. AWS AI/ML services are deployed across multiple Availability Zones for resiliency, fault tolerance, and scalability. Here are some design principles for reliability:

- To evaluate performance and test model rollback, enable a deployment testing strategy.

- Implement automation whenever possible to track changes. This can be achieved with an MLOps strategy and SageMaker Pipelines.

- Configure autoscaling for SageMaker endpoints and IUs for Rekognition Custom Labels and Lookout for Vision.

Performance efficiency

Performance efficiency focuses on maintaining efficient workloads, monitoring performance, and optimizing resource usage. Here are some best practices to consider:

- Understand the resource options available and evaluate the trade-offs of Rekognition, Lookout for Vision, and SageMaker for your CV workloads

- Monitor the health and performance of Rekognition and Lookout for Vision workloads using CloudWatch

- Monitor for data drift and model quality drift with SageMaker Clarify and SageMaker Model Monitor

- Automate retraining with MLOps

- Include human-in-the-loop to automate review workflows and evaluate low confidence scores using Amazon A2I

Security

Security focuses on protecting data and infrastructure, limiting user permissions, and implementing an incident response strategy. This also includes establishing governance, which we will discuss in the next chapter. Here are some design principles for security:

- Encrypt data at rest and in transit. Rekognition and Lookout for Vision encrypt data at rest, which is stored in S3 buckets, and use AWS KMS keys for server-side encryption. The API endpoints are encrypted via TLS.

- Enforce least privilege access, develop a multi-account strategy to separate workloads, and implement guard rails.

- Log activity and API calls with AWS CloudTrail to detect suspicious activity.

- Use Amazon VPC endpoints to privately connect your VPC and AWS AI/ML services without accessing the public internet.

Sustainability

Sustainability focuses on improving the environmental impact by minimizing consumption and increasing the efficiency of workloads. Here are some best practices for sustainability:

- Use AWS AI services and pre-trained models to decrease the resources that are required to deploy an ML solution.

- Implement autoscaling and shut down idle resources to decrease utilization.

- Optimize deployment compute resources and implement benchmarking. Consider SageMaker Serverless Inference or other inference options.

In this section, we covered the six pillars of the AWS Well-Architected Framework. We included best practices to incorporate into your CV workloads to optimize cost, secure infrastructure, monitor performance, and implement reliability.

Summary

In this chapter, we provided best practices and tips for designing an end-to-end CV pipeline. We discussed the importance of defining a problem that CV can solve before moving toward the data collection stage of your project. In addition, we highlighted techniques for preprocessing data, training, evaluating, tuning, deploying, and monitoring a model. Next, we covered how to develop an MLOps strategy. Lastly, we summarized the AWS Well-Architected Framework and addressed considerations for architecting secure, scalable, reliable, and efficient CV workloads. In the next chapter, we will define AI governance and detail how to establish a framework for applying governance to your CV workloads.

13
Applying AI Governance in CV

In the previous chapter, we covered best practices for designing an end-to-end CV pipeline. We discussed how you can use these guidelines throughout the ML life cycle to build, deploy, and manage reliable and scalable CV workflows.

In this chapter, we will discuss AI governance and its importance in CV. You may be asking, "How is AI governance relevant to my role as an AI/ML practitioner?" Security and compliance are only a small facet of the components of an AI governance strategy. The lack of existence of an organizational AI governance strategy has implications across the entire ML life cycle. From data collection to deploying and monitoring models, as an AI/ML practitioner it's your responsibility to work with other business stakeholders to ensure ML models are performing as expected, and to address problems that arise quickly and efficiently.

The increasing speed and scale of model development and deployment has created new sets of challenges. Many tools and applications are available to assist with the technical processes of designing and implementing production-ready ML models. However, few tools and guidelines exist that define the business processes that should be considered when developing an AI governance framework. Establishing a repeatable set of business processes and procedures is essential to ensure that your organization's ML projects continue to deliver impactful results. We will dive deeper into these mechanisms and tasks to create a secure and resilient AI system infrastructure.

In this chapter, we will cover the following:

- Understanding AI governance
- Applying AI governance in CV
- Using Amazon SageMaker for governance

Understanding AI governance

AI governance is the process of implementing controls and establishing processes and procedures to minimize risks of the development and use of AI systems, while maximizing the delivery of business outcomes. AI systems are prone to the same risks as other technology stacks. Their infrastructure needs to be designed to be resilient, scalable, and be able to withstand security vulnerabilities and cyber attacks. They also face additional unique risks, which we will discuss throughout this section.

ML models require large amounts of data. The amount of data that is available for ML development is rapidly increasing, and not all of this data is beneficial or relevant for solving an AI/ML problem. There is potential for unintended bias within the data processing phase and throughout the entire ML life cycle. Organizations also have to abide by regulations and compliance standards. By proactively establishing an organizational AI governance framework, you can identify potential problems early on, implement effective processes, and continue to successfully deliver meaningful insights and value from your ML models. The following figure shows the stages of the ML life cycle, with key components responsible stakeholders should consider for an AI governance framework. The key components listed are not an all-inclusive list of factors to consider but are a good starting point for providing oversight of AI system infrastructure.

ML life cycle stage	Key components	Business stakeholders
Define business problem and objectives	Define technical requirements, documentation, risks, compliance rules	Business analyst, domain expert, organizational leaders, AI/ML architect, auditor, compliance expert
Data processing	Clean data, validate quality, understand structure, detect bias	Data architect, data engineer, domain expert, data scientist, AI/ML architect
Model development	Auditability, lineage tracking, versioning	Data scientist, ML engineer, domain expert, AI/ML architect, MLOps engineer, security engineer
Model deployment	Automate model deployment, explainability and interpretability, visibility, verify compliance	Domain expert, MLOps engineer, model risk manager, AI/ML architect, security engineer
Model monitoring	Establish a monitoring system to detect risks, data drift, model quality drift, promote fairness	Business analyst, domain expert, organizational leaders, AI/ML architect, auditor, compliance expert, model risk manager, security engineer, MLOps engineer

Figure 13.1 – AI governance across the ML life cycle

Defining risks, documentation, and compliance

With any risk framework, it is important to identify the potential risks and the likelihood of them occurring to design a mitigation strategy. Potential risks depend upon an organization's current risk infrastructure and differ between industries. You should also understand the current risk infrastructure and controls that your organization has in place to effectively manage and measure AI risks. Also, it is key that teams maintain documentation for the entire AI system to promote communication across teams and to understand the impacts of technology decisions. Practices and procedures should be put in place for testing the resiliency and security of the system, prioritizing imminent risks, and incorporating stakeholder feedback. AI governance processes should be reviewed regularly and adapted to the organization's current level of risk tolerance. Across industries, there are regulations and compliance standards designed to promote data privacy and protect users. Discussing the details of current regulatory and legal frameworks is outside the scope of this book, but it is crucial to understand the implications of non-compliance and how to design AI systems that achieve compliance standards. Organizations that are found to be non-compliant can face serious legal ramifications and reputational risk.

Data risks and detecting bias

One important risk category within AI/ML is related to data. AI systems are only as valuable in creating meaningful predictions and delivering business value as the quality of the data used. Poor quality data not only results in incorrect predictions but may lead to unintended consequences such as potential bias and data privacy issues. Biased data, or data that is not utilized properly, can lead to discrimination and unfair outcomes. One technique to decrease the amount of biased data that an AI system uses is to clean the data and incorporate explainability techniques. We will cover more techniques to help manage data bias and discuss the impact of unfairly biased data in CV in the next section.

Organizations frequently use AI systems for business-critical decision-making processes. Not having data governance processes and proper controls in place that provide data transparency may lead to data ethics concerns. Data governance is essential for analyzing privacy risks and protecting user data. If a data privacy attack occurs, the bad actor could gain access to the training data and compromise data privacy. Documenting metadata or using a feature store to track data lineage ensures that data consistency and monitoring are maintained across AI/ML pipelines. Establishing this consistency helps automate the ML development and deployment process. By decreasing the number of manual processes in place, organizations can make better use of systems that can monitor and manage the risks of AI systems.

Auditing, traceability, and versioning

Audits are often performed to meet regulatory compliance requirements. Internal security audits should also be implemented to assess and test systems for security risks. Enabling versioning of ML artifacts helps to promote the transparency and trustworthiness of AI systems. Performing versioning and experiment tracking allows for proper auditing of all artifacts and a better understanding of key metrics. Model lineage is also useful for gaining visibility into the AI/ML workflow, such as understanding access controls, tracking data lineage, and identifying risks. Using a model registry is beneficial for cataloging and managing model artifacts, tracking versions, and automating production model deployment.

Monitoring and visibility

An effective AI governance framework assures the monitoring and visibility of the activities of an AI system. Visibility is needed to quickly detect potential problems early. A monitoring system should be enabled for alerting, identifying risks, enhanced explainability and interpretability, detecting model quality and data drift, and maintaining granular access control. As an AI system continues to receive new inputs of data, the quality of the data might shift or the data that the model was previously trained on may differ greatly from the new data. This can result in model accuracy drift or data drift. Drift can lead to inherent risks. Alerting on data drift early helps mitigate biased outcomes and security risks. If the model experiences accuracy drift, the underlying issue might be attributed to a data poisoning attack where the training data is contaminated. A data drift alert may indicate that the training data was unfairly biased. Monitoring detects these types of drift early so that the model can be retrained before business outcomes are significantly impacted.

Interpretability gives insights into a model's behavior and how it generates predictions. Models are increasing in complexity. Interpretability promotes transparency and provides business stakeholders with an understanding of how a model's inferences are contributing to business outcomes. Interpretability is also sometimes needed for regulatory requirements. Explainability, often referred to as explainable AI, is similar to interpretability. Its purpose is to explain a model's behavior in terms humans can understand, but it also details why a decision was made. Both interpretability and explainability are frequently needed to meet regulatory requirements. There are several interpretability and explainability AI methods that we will discuss in more detail later in this chapter using Amazon SageMaker Clarify.

It is important to provide least-privilege access to an AI system and to monitor permissions. If unauthorized access is granted or the system is breached, a monitoring system should trigger an alert for remediation. For auditing purposes and to monitor activity, enable data access logging. Data access logging monitors data access patterns and assists in alerting on suspicious or anomalous activity.

MLOps

In *Chapter 12*, we discussed MLOps fundamentals and the services that are available in AWS for implementation. Tasks that are committed manually not only slow down development time but are prone to errors. A **continuous integration and continuous development** (**CI/CD**) pipeline automates repeatable tasks in the **software development life cycle** (**SDLC**). This helps speed up the development and delivery of applications. The same repeatable mechanism can be adopted for automating an end-to-end ML workflow. Automating the steps to build, deploy, and manage models helps promote governance across the ML life cycle.

Responsibilities of business stakeholders

AI governance requires cross-collaboration and open communication among business stakeholders. Stakeholders need to be aligned on their roles and responsibilities, since many responsibilities overlap across the ML life cycle. They also should automate repeatable processes to reduce overhead and quickly address risks to the AI system infrastructure. Many organizations create a **Center of Excellence (CoE)** to ensure key stakeholders are represented across the organization, to regularly review the AI governance framework, and to share best practices and procedures. The following is a list of stakeholders often involved across the ML life cycle and their responsibilities:

- **Business analysts** – Understand the AI project and the desired business outcomes, and define what tasks to focus on that deliver value.

- **Organizational leaders** – Need to understand how the model works and its limitations. They also establish how the model will be used in the organization.

- **Domain experts** – Have deep knowledge and understanding of the ML problem. They work closely with data scientists and ML engineers. Some of the questions they ask are, "Is the model using the right features for the use case?" and "Is the right data being included in the training dataset?"

- **Data architects and data engineers** – Choose the data sources for the AI system and are responsible for ingesting and transforming the data for the ML model.

- **Data scientists** – Develop the ML model, understand how the model works, and evaluate it after training to improve its performance.

- **AI/ML architects** – Design the architecture of the AI system and are responsible for choosing the right technical components for tasks across the entire ML life cycle.

- **ML engineers** – Work with the data scientists to design and implement an AI system for production.

- **MLOps engineers** – Build and orchestrate the automation of an end-to-end ML pipeline and incorporate governance components such as monitoring and traceability.

- **Security engineers** – Manage user access permissions and set up security guardrails for the AI system. Also, they set up audit trails and determine who has access to the data for training the model.

- **Model risk managers** – Evaluate the risk of the model and ensure that requirements such as model versioning, explainability, and monitoring are enabled for effective oversight.

- **Auditor and compliance experts** – Oversee that the AI system is meeting compliance and legal requirements. They review that the model is transparent, fair, reliable, and safe for consumers.

In this section, we discussed the meaning of AI governance and summarized the processes to consider to establish an AI governance framework. We also covered the roles and responsibilities of business stakeholders. In the next section, we will discuss how governance is applied to CV and dive deep into addressing bias.

Applying AI governance in CV

One example CV use case is automating a facial identification workflow. Manually verifying a person's identity is a process that is time-consuming, inaccurate, and difficult to scale. Using biometrics and CV algorithms for identity verification can reduce friction during the customer onboarding process and lead to a better customer experience. Another application of CV is in fraud detection. Identifying anomalies in data helps catch and prevent fraudulent transactions. However, there are additional risks involved in automating facial recognition and fraud prevention workflows. Biased datasets could lead to models making predictions based on gender, income, ethnicity, age, and other discriminatory characteristics. To help mitigate these added risks and promote algorithmic fairness, it is important to have an ethical AI governance model that continuously reinforces transparency and the decision-making processes of AI systems.

Types of biases

Building fair algorithms is challenging and human biases exist throughout the entire ML life cycle, including during data collection, annotation, and modeling. Identifying the types of biases is often difficult to recognize, due to the large size of datasets and the complexity of models. Before we discuss techniques to deal with bias, let's define some types of biases:

- **Selection bias** – This occurs when there is a preference for certain types of data and the samples are not random. For example, when training a facial recognition model, the images used are heavily representative of one race. The model's predictions would then be biased and inaccurate when testing images of other ethnicities.

- **Label bias** – Humans assign different labels to objects in images. In the following image, they might assign the labels "leaves" versus "leaf" versus "cucumber leaf."

Figure 13.2: Leaf sample

- **Algorithmic bias** – Stereotypes or prejudice of the humans involved unfairly skew the training data and results of the model.

- **Evaluation bias** – The metrics used to evaluate the model are not representative of the use case, such as evaluating the accuracy of a facial recognition model by only including images from a small subset of age groups or skin colors. The benchmarks for evaluation should be impartial and the validation dataset should effectively represent the entire use case.

Mitigating bias in identity verification workflows

AI governance helps mitigate and minimize the risk of bias. Governance throughout the ML life cycle provides a standardized set of processes and tools for organizations to detect and address bias. Let's summarize how to incorporate governance in an identity verification workflow.

Data collection and processing

Facial images are used for identity verification models and these types of images require that consent and authorization be obtained for their permissible use. Therefore, you should understand the laws and regulations regarding data use and retention. Enable encryption to protect data from unauthorized use and set retention policies to determine when and how long images can be used. To assist with understanding what data sources the images are being obtained from for training, enable traceability and transparency. When collecting image data, consider their original source such as if they are from an open source repository, understand the privacy and licensing considerations to follow, and perform exploratory data analysis. Exploratory data analysis entails analyzing the dataset for completeness, understanding its main characteristics, reviewing the accuracy of annotation samples, and assessing whether the data is diverse and representative of the entire population. Generating visual diagrams such as histograms and summary plots is helpful for visualizing features, detecting patterns in the data, and finding anomalies. This is also useful for addressing class imbalance – for example, identifying that males are represented in the training images at a much higher percentage compared to females. One method to measure class imbalance and mitigate pre-training bias is to use SageMaker Clarify. In the next section, we will cover SageMaker Clarify in more detail.

Building and evaluating a model

All models are inherently biased. The goal of an AI system is to achieve fairness – producing similar outcomes for similar groups or populations – by reducing unwanted bias. If possible, choose explainable algorithms that give details about their design and the rationale behind their predictions. Explainability can also be used to understand the model's behavior and for evaluating post-training bias metrics. In an identity verification model, the importance of input features on the model's predictions can be explained using **SHapley Additive exPlanations (SHAP)** values. SHAP values are derived from game theory. The mathematics behind their calculation is complex and outside the scope of this book; they are also useful for explaining the correlation between features. Another technique that is useful for explaining image classification models is **local interpretable model-agnostic explanations (LIME)**.

LIME is similar to SHAP. It also evaluates the contribution of features to individual predictions, but it lacks some of the additional features of SHAP. Including explainability techniques helps establish confidence in the identity verification model's predictions. Even if the predictions meet established confidence thresholds, human review or oversight is often needed for testing. Human reviewers should have domain expertise, recognize their own biases, and oversee that the model is being used as intended.

Monitoring and continuous improvement

Once the model is deployed, it needs to continue to be monitored for any changes in performance and model drift. Amazon SageMaker Model Monitor can be used to monitor different types of drift such as data quality, model quality, bias drift, and feature attribution drift. Reports can also be generated for visual analysis to assist with understanding the model's behavior. If an identity verification model decreases in accuracy, this can have unintended consequences and serious implications. This could result in discrimination from its predictions and loss of user trust. Establishing the proper guardrails, such as security controls to mitigate security and privacy risks, regularly testing the model's accuracy, using humans to review predictions, and monitoring for bias is necessary for maintaining robustness. As input data consistently changes and the use case evolves, include feedback mechanisms for users and stakeholders to help continuously improve the system.

In this section, we discussed how AI governance can be applied in CV. We defined the types of biases that exist and outlined the tasks to help mitigate bias in identity verification workflows. In the next section, we will discuss how to use Amazon SageMaker to promote governance.

Using Amazon SageMaker for governance

Throughout this chapter, we have detailed the importance of establishing an AI governance framework. However, setting up an overall process to gain visibility of performance, control access, audit changes, and mitigate bias is no easy feat. To help address these challenges and remove undifferentiated heavy lifting, SageMaker provides purpose-built tools to help implement governance.

ML governance capabilities with Amazon SageMaker

As user adoption increases, it becomes more difficult for administrators to manage user access to ML projects. Custom permissions policies are often required for different ML user groups and these permissions sets vary greatly. Customization is a time-consuming process that could delay user onboarding. **Amazon SageMaker Role Manager** (https://docs.aws.amazon.com/sagemaker/latest/dg/role-manager.html) simplifies this process by providing a baseline set of permissions for different user personas and ML activities through IAM policies. There are predefined ML activities for MLOps, data scientists, and the SageMaker compute personas; these can also be customized. You also have the ability to customize network access and define encryption keys.

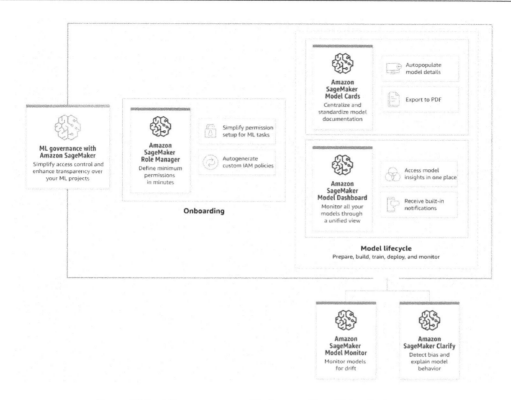

Figure 13.3 – ML governance with Amazon SageMaker features

As the number of models increases within your organization, another challenge is capturing the required information and standardizing documentation for auditing purposes and explaining model behavior. Many ML practitioners use disparate tools to gather this information. **Amazon SageMaker Model Cards** (https://docs.aws.amazon.com/sagemaker/latest/dg/model-cards.html) auto-populates your model information, including training details, and stores this information in a central repository in the SageMaker console. You can also add details such as the model's intended use and training observations, and attach model evaluation results. Model cards can be exported and shared with business stakeholders or your customers. This helps standardize your model documentation and allows you to more easily meet your compliance requirements.

Gaining visibility of a model's performance is another challenge and most organizations often do not have a consolidated view of all their models. This makes it more difficult to understand when a model is deviating from normal behavior and to set up automated alerting. **Amazon SageMaker Model Dashboard** (https://docs.aws.amazon.com/sagemaker/latest/dg/modeldashboard.html) provides a unified view for tracking deployed models and model behavior violations. It is integrated with SageMaker Model Monitor and SageMaker Clarify to alert on drift and errant model monitoring jobs. This allows you to quickly troubleshoot model performance and take corrective actions.

Amazon SageMaker Clarify for explainable AI

We have discussed that bias can appear throughout the entire end-to-end ML workflow. Bias can show up in training datasets, after training when making predictions, and over time as drift occurs with changes to the data. With all these different types of biases that could exist, it is difficult to monitor and get insights across each stage of the ML life cycle. SageMaker Clarify is one solution that supports the transparency and explainability of ML workloads. It is integrated with several other Amazon SageMaker solutions such as Amazon SageMaker Data Wrangler to identify class imbalances in data during data preparation, and SageMaker Model Monitor, which works with Amazon CloudWatch to alert for changes in model bias and detect data quality and model quality drift. SageMaker Clarify is also integrated with Amazon SageMaker Experiments to provide visualizations and graphs for transparency that show which features had the most influence on the model.

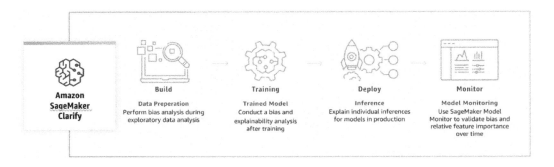

Figure 13.4 – Features of SageMaker Clarify across the ML life cycle

SageMaker Clarify is beneficial for many scenarios where model explainability is needed or even required, such as regulatory compliance, internal auditing and reporting, and providing insights into the results of a model's decisions to end users. It also provides explanations of a model's predictions in near real time, which is useful for customer service use cases where problems need to be quickly resolved. Earlier in this chapter, we discussed the feature attribution method called Shapley values, which assigns each feature a value of importance for a model's prediction. SageMaker Clarify uses this method for explainability and can be used to gain insights into your CV models (`https://docs.aws.amazon.com/sagemaker/latest/dg/clarify-model-explainability-computer-vision.html`).

In this section, we covered how to improve governance with Amazon SageMaker. We detailed how to generate customized roles for different user personas using SageMaker Role Manager, how to streamline model documentation with SageMaker Model Cards, and how to get a comprehensive view of model performance with SageMaker Model Dashboard. Also, we discussed using SageMaker Clarify to help explain your model's predictions.

Summary

In this chapter, we covered establishing an AI governance framework. We discussed processes and procedures to help minimize risk and maximize the results of AI/ML systems. Next, we defined the key roles and responsibilities of business stakeholders for effective governance. We also summarized how AI governance applies to CV and how to mitigate unfair bias. Lastly, we detailed how to use Amazon SageMaker to apply governance and explainability across your workloads.

In this book, we've taken a journey into exploring the world of CV. From simple image classification tasks to the recent excitement surrounding generative AI, we've only just begun to discover the applications of CV. We hope you've enjoyed walking through some real-world examples and learning tips and tricks for using Amazon Rekognition, Amazon Lookout for Vision, and Amazon SageMaker along the way. Now, you're ready to solve your business challenges with CV and incorporate AWS AI/ML services into your next project.

Index

Packtpub.com

Subscribe to our online digital library for full access to over 7,000 books and videos, as well as industry leading tools to help you plan your personal development and advance your career. For more information, please visit our website.

Why subscribe?

- Spend less time learning and more time coding with practical eBooks and Videos from over 4,000 industry professionals

- Improve your learning with Skill Plans built especially for you

- Get a free eBook or video every month

- Fully searchable for easy access to vital information

- Copy and paste, print, and bookmark content

Did you know that Packt offers eBook versions of every book published, with PDF and ePub files available? You can upgrade to the eBook version at packtpub.com and as a print book customer, you are entitled to a discount on the eBook copy. Get in touch with us at customercare@packtpub.com for more details.

At www.packtpub.com, you can also read a collection of free technical articles, sign up for a range of free newsletters, and receive exclusive discounts and offers on Packt books and eBooks.

Other Books You May Enjoy

If you enjoyed this book, you may be interested in these other books by Packt:

Applied Geospatial Data Science with Python

David S. Jordan

ISBN: 9781803238128

- Understand the fundamentals needed to work with geospatial data
- Transition from tabular to geo-enabled data in your workflows
- Develop an introductory portfolio of spatial data science work using Python
- Gain hands-on skills with case studies relevant to different industries
- Discover best practices focusing on geospatial data to bring a positive change in your environment
- Explore solving use cases, such as traveling salesperson and vehicle routing problems

3D Deep Learning with Python

Xudong Ma, Vishakh Hegde, Lilit Yolyan

ISBN: 9781803247823

- Develop 3D computer vision models for interacting with the environment
- Get to grips with 3D data handling with point clouds, meshes, ply, and obj file format
- Work with 3D geometry, camera models, and coordination and convert between them
- Understand concepts of rendering, shading, and more with ease
- Implement differential rendering for many 3D deep learning models
- Advanced state-of-the-art 3D deep learning models like Nerf, synsin, mesh RCNN

Packt is searching for authors like you

If you're interested in becoming an author for Packt, please visit authors.packtpub.com and apply today. We have worked with thousands of developers and tech professionals, just like you, to help them share their insight with the global tech community. You can make a general application, apply for a specific hot topic that we are recruiting an author for, or submit your own idea.

Share Your Thoughts

Now you've finished *Computer Vision on AWS*, we'd love to hear your thoughts! Scan the QR code below to go straight to the Amazon review page for this book and share your feedback or leave a review on the site that you purchased it from.

https://packt.link/r/1-801-07868-8

Your review is important to us and the tech community and will help us make sure we're delivering excellent quality content.

Download a free PDF copy of this book

Thanks for purchasing this book!

Do you like to read on the go but are unable to carry your print books everywhere?

Is your eBook purchase not compatible with the device of your choice?

Don't worry, now with every Packt book you get a DRM-free PDF version of that book at no cost.

Read anywhere, any place, on any device. Search, copy, and paste code from your favorite technical books directly into your application.

The perks don't stop there, you can get exclusive access to discounts, newsletters, and great free content in your inbox daily

Follow these simple steps to get the benefits:

1. Scan the QR code or visit the link below

https://packt.link/free-ebook/9781801078689

2. Submit your proof of purchase
3. That's it! We'll send your free PDF and other benefits to your email directly

www.ingramcontent.com/pod-product-compliance
Lightning Source LLC
Chambersburg PA
CBHW080624060326
40690CB00021B/4801